Developing Formative Assessment in STEM Classrooms

This book, drawn from the practical experience of the European Assess Inquiry in Science, Technology and Mathematics Education (ASSIST-ME) project, aims to support researchers, teacher educators and teachers in navigating the complex nature of formative assessment.

It delves into how formative assessment methods may adapt different features within STEM classroom practices and how formative and summative assessment could be coordinated. The book illustrates how ideas about inquiry and assessment are played out in the classroom, revealing both how the pedagogical approach fits with the assessment choices made and the support that teachers need, to engage with these changes. It informs and presents some of the challenges related to the uptake and development of inquiry learning in STEM classrooms. The chapters in this volume address three questions:

- What formative assessment objectives and methods are relevant and appropriate in science education?
- How do teachers monitor student learning as it unfolds during inquiry-based teaching–learning sequences?
- To what extent does a partnership between researchers and teachers make it possible to implement effective assessment methods and acquire relevant professional knowledge?

Taken together, they explore how national educational systems can foster and build the systemic support measures and tools that teachers need to integrate into their classroom practices that fit alongside or within their summative assessment practices. The chapters included in this book were originally published in the *International Journal of Science Education*.

Michel Grangeat is Professor Emeritus of Educational Sciences at University of Grenoble Alpes, France. Michel's research focuses on the activities of professionals in the field of human relations and in collective contexts. The activities studied are those aimed at improving the well-being, development and learning of individuals, whether adults or children.

Christine Harrison is Professor of Science Education at King's College London, UK. Her research focuses on teachers' conceptualization of classroom assessment and particularly how assessment for learning pedagogy is implemented in STEM classrooms. Her research studies enable her to work alongside teachers through a close-to-practice approach that both learns from and informs practice.

Jens Dolin is Professor Emeritus of Science Education at University of Copenhagen, Denmark. He has a broad research portfolio in the learning and teaching of science with foci on development of students' science competences and bildung, especially through cross-curricular education. Assessment aspects and involving policy stakeholders in the educational processes have been central aspects of the research.

Developing Formative Assessment in STEM Classrooms

Edited by
Michel Grangeat, Christine Harrison and Jens Dolin

Routledge
Taylor & Francis Group
LONDON AND NEW YORK

First published 2024
by Routledge
4 Park Square, Milton Park, Abingdon, Oxon OX14 4RN

and by Routledge
605 Third Avenue, New York, NY 10158

Routledge is an imprint of the Taylor & Francis Group, an informa business

British Library Cataloguing in Publication Data
A catalogue record for this book is available from the British Library

ISBN13: 978-1-032-73799-7 (hbk)
ISBN13: 978-1-032-73817-8 (pbk)
ISBN13: 978-1-003-46607-9 (ebk)

DOI: 10.4324/9781003466079

Typeset in Minion Pro
by Newgen Publishing UK

Publisher's Note
The publisher accepts responsibility for any inconsistencies that may have arisen during the conversion of this book from journal articles to book chapters, namely the inclusion of journal terminology.

Contents

Citation Information

The following chapters, except for Chapter 3, were originally published in the *International Journal of Science Education*, volume 43, issue 3 (2021). Chapter 3 was originally published in volume 41, issue 15 (2019) of the same journal. When citing this material, please use the original page numbering for each article, as follows:

Introduction

Exploring assessment in STEM inquiry learning classrooms
Michel Grangeat, Christine Harrison and Jens Dolin
International Journal of Science Education, volume 43, issue 3 (2021), pp. 345–361

Chapter 1

Forms and functions of on-the-fly formative assessment conversations in physics inquiry lessons
Pasi Nieminen, Markus Hähkiöniemi and Jouni Viiri
International Journal of Science Education, volume 43, issue 3 (2021), pp. 362–384

Chapter 2

French teachers' informal formative assessment in the context of inquiry-based learning
Elie Rached and Michel Grangeat
International Journal of Science Education, volume 43, issue 3 (2021), pp. 385–406

Chapter 3

Making learning effective – quantity and quality of pre-service teachers' feedback
Mathias Ropohl and Silke Rönnebeck
International Journal of Science Education, volume 41, issue 15 (2019), pp. 2156–2176

Chapter 4

Analysing formal formative assessment activities in the context of inquiry at primary and upper secondary school in Switzerland
Regula Grob, Monika Holmeier and Peter Labudde
International Journal of Science Education, volume 43, issue 3 (2021), pp. 407–427

Chapter 5

Development of teachers' perspective on formative peer assessment
Iva Zlabkova, Jan Petr, Iva Stuchlikova, Lukas Rokos and Alena Hospesova
International Journal of Science Education, volume 43, issue 3 (2021), pp. 428–448

Chapter 6

The what, when & how factors: reflections on classroom assessment in the service of inquiry
Bronwen Cowie and Christine Harrison
International Journal of Science Education, volume 43, issue 3 (2021), pp. 449–465

For any permission-related enquiries please visit:
www.tandfonline.com/page/help/permissions

Notes on Contributors

Ron Blonder, Department of Science Teaching, Weizmann Institute of Science, Israel.

Bronwen Cowie, Division of Education, The University of Waikato, Hamilton, New Zealand.

Jens Dolin, Department of Science Education, University of Copenhagen, Copenhagen, Denmark.

Michel Grangeat, Department of Educational Science, University of Grenoble Alpes (LaRAC), Grenoble, France.

Regula Grob, University of Teacher Education (UTE) Lucerne, Switzerland.

Markus Hähkiöniemi, Department of Teacher Education, University of Jyväskylä, Jyväskylä, Finland.

Christine Harrison, School of Education, Communication and Society, King's College, London, UK.

Monika Holmeier, Teaching and Curriculum Development, University of Applied Science and Arts Northwestern Switzerland, Brugg-Windisch, Switzerland.

Alena Hospesova, Department of Primary and Pre-school Education, Faculty of Education, University of South Bohemia in Ceske Budejovice, Ceske Budejovice, Czech Republic.

M. Gail Jones, College of Education, North Carolina State University, Raleigh, USA.

Peter Labudde, Centre for Science and Technology Education, University of Applied Science and Arts Northwestern Switzerland, Muttenz, Switzerland.

Pasi Nieminen, Department of Teacher Education, University of Jyväskylä, Jyväskylä, Finland.

Jan Petr, Department of Biology, Faculty of Education, University of South Bohemia in Ceske Budejovice, Ceske Budejovice, Czech Republic.

Elie Rached, Studies in Science and Technology Laboratory, Paris-Saclay University, France.

Lukas Rokos, Department of Biology, Faculty of Education, University of South Bohemia in Ceske Budejovice, Ceske Budejovice, Czech Republic.

Silke Rönnebeck, Department of Chemistry Education, IPN Leibniz-Institute for Science and Mathematics Education, Kiel, Germany.

Mathias Ropohl, Chemistry Education, University of Duisburg-Essen, Essen, Germany.

Iva Stuchlikova, Department of Psychology, Faculty of Education, University of South Bohemia in Ceske Budejovice, Ceske Budejovice, Czech Republic.

Jouni Viiri, Department of Teacher Education, University of Jyväskylä, Jyväskylä, Finland.

Iva Zlabkova, Department of Education, Faculty of Education, University of South Bohemia in Ceske Budejovice, Ceske Budejovice, Czech Republic.

Foreword

Ron Blonder and M. Gail Jones

The book, *Developing Formative Assessment in STEM Classrooms*, pushes research on formative assessment in science, technology, engineering and mathematics (STEM) education in new directions. The studies originate from a special issue of the *International Journal of Science Education* (*IJSE*) published in 2021 that focuses on assessment within STEM inquiry. These papers afford science educators the opportunity to reexamine a pivotal subject within the teaching and learning process.

Assessment is critically important in shaping and supporting learning and it is particularly influential when assessment addresses epistemic understanding and skills that are difficult to assess using conventional means such as testing. This topic is of international interest, as evidenced by the international collaboration of researchers contributing chapters from various countries.

The questions of what to assess, when and how are discussed in the study by Cowie and Harrison (2021), and the remaining chapters in the book delve into research focusing on models of formative assessment. It's not surprising that formative assessment is an essential component of student-centered pedagogies. Formative assessment within the context of inquiry-based is discussed in papers by Cowie and Harrison (2021), Grangeat et al. (2021), Grob et al. (2021), and Rached and Grangeat (2021). As these chapters reveal, applying formative assessment during inquiry-based learning supports the dynamic development of student skills at multiple levels of science education. The papers in the book link formative assessment to other student-centered pedagogies, e.g. project-based learning (King-Dow, 2022).

Implementing formative assessment effectively is a challenge for educators, but the process of developing and reflecting on learning in action can support professional development (Blonder, 2018; Grangeat et al., 2021). Making formative assessment a part of teacher education from the beginning is important for teacher development (Ropohl & Rönnebeck, 2019) by preparing teachers to become effective in using peer assessment (Zlabkova et al., 2021) and to develop strategies for providing feedback during class conversation (Nieminen et al., 2021). Grangeat et al. (2021) describe the role of supporting teachers as they develop both assessment and inquiry instructional skills.

These studies have served as the foundation for ongoing research that has explored STEM teachers' perception and their actual implementation of different assessment methods (Akiri et al., 2021) and also for the development of assessment literacy among educators (Hung & Wu, 2023).

Recognizing the significance of assessment within the context of learning became particularly salient amid the backdrop of the COVID-19 pandemic. Educators found themselves geographically separated from their students, prompting a pressing need for instructional evidence and tools enabling the ongoing monitoring of student comprehension during remote learning. However, as we transition back to traditional classroom settings, it has become evident that the mere physical presence of students in the classroom does not guarantee effective learning outcomes. Educators and researchers have come to appreciate that, devoid of appropriate knowledge and assessment tools to gauge skills, knowledge acquisition and other learning objectives during instruction, effective teaching remains elusive.

Clearly, formative assessment provides instructors with crucial insights throughout the learning process. This information serves as a foundation for adaptive teaching, wherein the educator tailors their pedagogical approach in dynamic instructional contexts. Given the contemporary focus on these themes within the realm of science education research, this book makes a noteworthy contribution to the ongoing scholarly discourse in this field.

References

Akiri, E., Tor, H. M., & Dori, Y. J. (2021). Teaching and assessment methods: STEM teachers' perceptions and implementation. *EURASIA Journal of Mathematics, Science and Technology Education*, *17*(6), em1969. https://doi.org/10.29333/ejmste/10882

Blonder, R. (2018). Student-curated exhibitions: Alternative assessment in chemistry education in Israel. In *International Perspectives on Chemistry Education Research and Practice* (Vol. 1293, pp. 39–55). Washington, DC: American Chemical Society. https://doi.org/doi:10.1021/bk-2018-1293.ch004

Cowie, B., & Harrison, C. (2021). The what, when & how factors: Reflections on classroom assessment in the service of inquiry. *International Journal of Science Education*, *43*(3), 449–465. https://doi.org/10.1080/09500693.2020.1824088

Grangeat, M., Harrison, C., & Dolin, J. (2021). Exploring assessment in STEM inquiry learning classrooms. *International Journal of Science Education*, *43*(3), 345–361. https://doi.org/10.1080/09500693.2021.1903617

Grob, R., Holmeier, M., & Labudde, P. (2021). Analysing formal formative assessment activities in the context of inquiry at primary and upper secondary school in Switzerland. *International Journal of Science Education*, *43*(3), 407–427. https://doi.org/10.1080/09500693.2019.1663453

Hung, C.-S., & Wu, H.-K. (2023). High school science teachers' assessment literacy for inquiry-based science instruction. *International Journal of Science Education*, 1–22. https://doi.org/10.1080/09500693.2023.2251657

King-Dow, S. (2022). The effects of cross-disciplinary life science innovation implemented by students' stimulated strategies for PBL-SELF efficacy. *Journal of Baltic Science Education*, *21*(6), 1069–1082. https://doi.org/10.33225/jbse/22.21.1069

Nieminen, P., Hähkiöniemi, M., & Viiri, J. (2021). Forms and functions of on-the-fly formative assessment conversations in physics inquiry lessons. *International Journal of Science Education*, *43*(3), 362–384. https://doi.org/10.1080/09500693.2020.1713417

Rached, E., & Grangeat, M. (2021). French teachers' informal formative assessment in the context of inquiry-based learning. *International Journal of Science Education*, *43*(3), 385–406. https://doi.org/10.1080/09500693.2020.1740818

Ropohl, M., & Rönnebeck, S. (2019). Making learning effective – Quantity and quality of pre-service teachers' feedback. *International Journal of Science Education*, *41*(15), 2156–2176. https://doi.org/10.1080/09500693.2019.1663452

Zlabkova, I., Petr, J., Stuchlikova, I., Rokos, L., & Hospesova, A. (2021). Development of teachers' perspective on formative peer assessment. *International Journal of Science Education*, *43*(3), 428–448. https://doi.org/10.1080/09500693.2020.1713418

Introduction: Exploring assessment in STEM inquiry learning classrooms

Michel Grangeat, Christine Harrison and Jens Dolin

ABSTRACT

This article introduces this special issue by addressing the types of assessment practices that are possible within STEM inquiry classrooms, where the focus is on noticing, developing, and improving inquiry competencies. Three domains are explored: informal formative assessment occurring in classroom interactions; structured formative assessment when teachers or peers collect information at a planned time; ongoing evolution of teachers' knowledge and practices. We analyse how classroom formative assessment practices develop alongside and support inquiry learning. We report on how STEM teachers might be supported in transforming their teaching practices and assessment methods in tandem. We focus on the role of classroom discussion to challenge students' ideas and make explicit their reasoning. We highlight how interactions between teachers and students support students in being more active in the assessment process. Combining inquiry learning with adapted assessment practices requires teachers to make significant changes to their practices. They need to be aware of how students develop their learning. Such professional moves are complex because implementing formative and summative assessments in a complementary manner remains a challenge for teachers and researchers. We emphasise how the collaboration between researchers and teachers support teachers in making changes to their professional knowledge, beliefs, and classroom practices.

Introduction

Assessment is one of the most important drivers in education and a defining aspect of any educational system. It involves the collection, interpretation and use of data for diverse purposes, since evidence for classroom assessment can be collected before, during, or after instruction (Zeng et al., 2018). Therefore, teachers can adopt different assessment methods ranging from formal, standardised procedures judged against predetermined criteria, to informal observations and discussions (Shavelson et al., 2008). Within this repertoire there are numerous assessment choices that teachers can make, depending on what purpose they intend to use the assessment information for (Bell & Cowie, 2001). In return, connecting assessment with specific teaching approaches, like inquiry learning in STEM education, contributes to fostering these teaching and learning methods (Harlen, 2009). In this article, we focus on the types of assessment practices that are possible within STEM inquiry classrooms, where the focus is on noticing. developing and improving inquiry competencies. Finding appropriate assessment methods for STEM inquiry classrooms is key in order for young people to be encouraged to aspire to careers in science, technology, engineering and mathematics and for policymakers to trust those assessments (Hazelkorn et al., 2015) and gain a better understanding of the processes involved.

Performing investigations in STEM education is to a large degree a holistic task where the various aspects of the inquiry process connect with and benefit from each other. Therefore, inquiry competences allow to act meaningfully in a complex situation by mobilising and involving different skills. For example, analysis of data can be influenced both by how accurately and skilled the student is in using a piece of apparatus and also by how systematic they have been in recording their data. Breaking inquiry down to separate skills in order to assess it more easily might therefore miss completely the purposes of the learning activity because students demonstrate their inquiry competences when they have to solve complex problems. As a result, using assessment protocols that simply check on separate skills in one context is insufficient to grasp an understanding of how learners undertake inquiry activities in STEM. Assessment of competences during inquiry activities will need more sociocultural oriented assessment formats than those for conceptual knowledge assessment (Dolin & Krogh, 2010). Moreover, one aspect that is key to all inquiry competency assessments is that assessment information can be collected in real time, as and when students are performing the inquiry, and, most importantly, as they are demonstrating their skills and competences, and revealing their thinking through the interactions they have with their fellow students and with their teacher (Matthews & McKenna, 2005). Nevertheless, assessing inquiry competences is also partly shaped by the inquiry approach and demand within that context, with sometimes, conceptual aspects being a key determinant, while, in other inquiries, procedural or technical knowledge play a more major role (Harrison, 2013). Accordingly, this article and this special issue focus on the respective roles of formative and summative assessment methods in STEM inquiry practices as they are part of a continuum that needs to be better explored and clarified; it does not focus on how to assess each competence specifically (i.e. assessing the argumentation competence or the modelling competence).

Putting into practice such an alignment between the purpose of the teaching (e.g. fostering inquiry skills or competences), the teaching methods and the assessment format is a challenge for STEM teachers. They are faced with decisions about when, what, and how to assess at various points in learning. These decisions and the resulting instructional practices are influenced by a number of factors linked with other classroom routines and practices, the educational context, and the quality of the interactions within the teacher community (Grangeat & Gray, 2007). Therefore, it is pertinent to explore the pedagogical and assessment trends associated with these approaches (Artigue & Blomhøj, 2013) and consider how teachers can be supported in building complementary instructional practices of inquiry and assessment. An effective way to help STEM teachers to adapt their teaching and assessment practices is through collaboration within a community of teachers, including facilitators such as researchers, and teacher educators

(Akerson et al., 2009). Such collaborative projects contribute to better harmonisation of inquiry-based science teaching and relevant assessment methods through researchers, teachers and teacher educators discussing and working on ideas of change together, within a supportive environment (Lepareur & Grangeat, 2018).

To support teachers across Europe in bringing about the radical change towards inquiry-based teaching methods suggested in the Rocard report (2007), there have been several European Union projects, such as MIND-THE-GAP, SAILS, S-TEAM, ESTABLISH, Fibonacci, PRIMAS and Pathway. What has been made clear, from these international projects, is that teacher education programmes need to be specifically designed and monitored for supporting inquiry teaching, as well as for prompting and developing an associated assessment shift (Hazelkorn et al., 2015; Heinz et al., 2017). These EU-projects have highlighted that changes in teaching methods towards more inquiry-oriented instruction are possible while connections between such an instruction and relevant assessment are more difficult to achieve.

This article and this special issue seek to elucidate which types of assessment tools, processes and approaches fit with inquiry-learning, in order to strengthen our understanding of the efficacy and practices of assessment that relates to inquiry-learning. This introductory article presents the background and research fields that are relevant to the two main questions that were the foci of the Assess Inquiry in Science, Technology and Mathematics Education (ASSIST-ME) project which the articles in this special issue address.

- How do teachers monitor student learning as it unfolds during inquiry-based teaching-learning sequences, using both informal and more structured assessment methods?
- How can a partnership between researchers and teachers be established for finding relevant ways for enacting effective assessment methods in inquiry classrooms and securing relevant professional knowledge?

These key questions are addressed in the papers in this special issue and emerge from close collaborative work between researchers and teachers, across eight countries. In the ASSIST-ME research project, both formative and summative assessment methods were developed and trialled to support and to improve inquiry-based approaches in European science, technology, and mathematics education, at both primary and secondary school level (Dolin & Evans, 2018).

Inquiry learning in STEM education

While much of the research on inquiry learning has been situated within science education, looking at education from a STEM perspective is relatively new, particularly for school systems in most European countries. STEM education is based on the idea of educating students in four specific disciplines – science, technology, engineering, and mathematics – in a cross-disciplinary and applied approach. Each of the subjects provides relevant starting points to introduce STEM thinking, reasoning, and problem-solving competences. Such links through different disciplines 'enable new ways of thinking and identifying solutions to problems that fall outside the boundaries of just one discipline' (Hazelkorn et al., 2015, p. 15).

STEM inquiry does not mean that all disciplines function along the same constructs and methods but rather share similar competences. While inquiry, as an approach to introducing critical thinking, is strongly related to science education, approaches to teaching and learning in mathematics, technology and engineering exist that share common goals and characteristics with inquiry-based science education (Greene & von Renesse, 2017; Lin et al., 2020). In mathematics, teachers tend to focus on comparing different ways of solving a problem and then debating, analysing effectiveness and efficacy and possibly reaching a consensus about fitness of method to context. Similarly, inquiry in technology and engineering focuses on comparing and contrasting design aspects, throughout the production process, leading to agreement and decision about quality. Common elements in all of these approaches consist of asking questions and testing out ideas in a systematic and iterative way, exploring evidence collected to justify or challenge initial ideas so that a consensus or negotiated ways forward can be reached and new knowledge developed.

Therefore, inquiry-based learning and teaching aim to support students in: building their understanding of fundamental scientific ideas; finding answers to their own and others' questions; exploring and using practices employed by scientists; taking charge of their learning as needed for genuine deep understanding.

Developing inquiry-based teaching practices

The articles in this special issue – as well as many EU-Projects – adopt the definition of STEM inquiry-learning by Linn et al. (2004, p. 4):

> [Inquiry is] the intentional process of diagnosing problems, critiquing experiments, and distinguishing alternatives, planning investigations, researching conjectures, searching for information, constructing models, debating with peers and forming coherent arguments.

Planning for inquiry requires teachers to focus on the process, contextual factors, and potential learning opportunities inherent in the activity, and, in addition, to demonstrate great flexibility both in what students do and how long different inquiry phases take. While there is a broad range of approaches to inquiry in classrooms, it is accepted that there are a number of teaching and learning elements that need to be present to distinguish it from more teacher-centred approaches (Grangeat, 2016; Minner et al., 2010). Depending on the particular field or topic, there are many and often diverse approaches to developing critical thinking through inquiry learning. Many teachers approach inquiry-learning in a practical way, encouraging students to collect data on phenomena and look for patterns in the evidence they collect, in order to work out relationships between variables, or map change over time. Other approaches range across researching ideas using internet resources to analyse and compare secondary data sets. Overcoming these variations, Rönnebeck et al. (2016) have developed a model for the concept of inquiry comprising three inquiry phases (preparing; carrying out; explaining and evaluating) that includes relating each phase to aspects of disciplinary knowledge. In this way, curriculum aims can be nested within inquiry learning activities and their respective assessment methods.

The aim in inquiry classrooms is to move away from set routines that allow students to simply witness phenomena towards involving students in developing a research question and in the process of exploring and building coherent explanations of phenomena in context (Kawalkar & Vijapurkar, 2013). While initially, the teacher might make decisions and direct students in what to do in an inquiry, as teachers and students grow in confidence with inquiry-learning, students begin to take on more active roles in deciding what to do and how to do it (Grangeat, 2015; Harrison, 2014; Van Driel & Berry, 2012). When ideas arise on how to answer an inquiry question, they are trialled, often with a range of methods attempted and adapted, as students problematise and work out ways forward that appear productive for their inquiry. This sequence of actions within inquiry is often accompanied, followed, or even primed by argumentation and debate with peers as the general competence of critical thinking through inquiry is both recognised and valued (Stewart et al., 2005; Windschitl et al., 2008).

Engaging students in inquiry-based learning

To engage students in inquiry learning, Barron and Darling-Hammond (2008) suggest that the problems to be investigated in an inquiry classroom need to be meaningful and realistic with multiple solutions and multiple methods for reaching solutions. This already sets demands on the teacher as they need to think through some of the possible solutions and methods, inherent within a specific inquiry, and how they would recognise if students were taking particular routes. This is no easy task because, for the students, these inquiry routes are new, and they may generate and trial several ideas before they hit on a particular idea or method that seems to have potential. While initial ideas and approaches might seem promising, trying to put these into action can cause students to reconsider their ideas, change tactics or adapt ideas. Equally, when they notice other groups of students approaching the inquiry task in a different way, this prompts the students to compare and analyse their initial choices and encourage them to question and compare their decisions with others.

As inquiry becomes a more shared and negotiated activity between the teacher and students, there is a consequential change in both their respective roles. Students become actively engaged in building their scientific thinking, and the teacher focuses on making students' understanding visible within classroom conversations (Furtak et al., 2016). Through discussion and debate with peers, students are encouraged to look for evidence that either supports or challenges their thinking (Chin, 2006; Harrison, 2013). This often involves teachers guiding activity through small group work and asking questions to encourage students to share ideas, promoting student thinking and enhancing self-reflection. In some instances, this requires the teacher to deliberately hold back a tendency to direct and lead student thinking, as student ideas unfold within conversations, only stepping in when wanting to probe student thinking more or when ideas move far off one of the tracks that the teacher hoped students would take. Furthermore, teachers might balance these more facilitative classroom practices with some direct teaching in that they 'also offer instruction in more traditional ways, such as lectures and explanations that are crafted and timed to support inquiry' (Barron & Darling-Hammond, 2008, p. 5).

For the teacher, making decisions based on evidence arising in the early stages of an inquiry is complex, because students could begin one line of inquiry and then jettison that one for another, before eventually deciding on a pathway to take (Biggers & Forbes, 2012). If a teacher samples the evidence too early, they may miss the key student ideas or evidence of misconceptions and yet, if they collect data too late, then they may miss crucial evidence of student thinking and current understanding. Such actions are therefore timely and will depend on judgements based on the incoming evidence from student inquiry and the consequential teacher response. As a result, it is difficult for teachers, during the lesson, to decide what type of assessment is most effective and when to conduct it, in the knowledge that sometimes the value of one choice over another is not realised until later in the inquiry process. Such choices could affect the reliability of the evidence as teachers refine their assessment capabilities within specific inquiries. To achieve inquiry-based learning in STEM subjects, the articles of this special issue explore to what extent formative assessment of student learning can be integrated into inquiry activities and how classroom assessment needs to be adapted at local level to ensure its effective use.

Assessment in STEM classrooms

Assessing goals is not easy in STEM subjects and the more complex the learning goals, the more difficult they are to assess. It is relatively simple to check on acquisition of STEM terminology, such as the names given to different structures, or whether a student can accurately use equipment or do a calculation. Conversely, assessing how conceptual ideas are put together and applied in an inquiry is more difficult. Moreover, there is general agreement in the literature that classroom and external assessments need to share the same or compatible underlying models of student learning (Krajcik et al., 2008; Pellegrino et al., 2001; Waddington et al., 2007).

For many educational systems, such coherence and alignment does not exist with the driver of assessment practices dominated by the external testing regime or the preponderance of summative purposes (Looney, 2011). In such educational systems, teachers need to navigate in two assessment systems that are not necessarily coordinated, one for formative purposes and the other for summative purposes. Although it is acknowledged that tensions will always exist between these two systems, some suggest moving away from the formative and summative dichotomy (Black & Wiliam, 2018; Harlen, 2009; Harrison, 2014). Dolin et al. (2018) investigate common grounds between formative and summative functions of assessment, which they rather consider as the two extremes of a broad continuum. In some cases, the same collected evidence could serve the two purposes on the condition that evidence elicitation is separated from its interpretation within the assessment process. Assessment data can then be used to inform on past learning, or to map present learning and drive next steps in learning.

Summative use of assessment is generally retrospective and has the purpose of summarising and reporting learning at a particular point. It involves reviewing learning over a period of time or checking-up by testing learning at a particular time and for that reason is also called 'assessment of learning'. The formative use of assessment assists learning during the learning process and for that reason is called 'assessment *for* learning' (Black et al., 2003; Wiliam, 2011). It involves processes of 'seeking and interpreting evidence for use by learners and their teachers to decide where the learners are in their learning and where they need to go and how best to get there' (ARG, 2002, p. 2). One assessment approach, with inquiry learning, is to assess mainly the written report or presentation that students produced on their investigation or project; a second approach for teachers is to look for assessment data during the inquiry activity itself, taking into account not only the written report but also student actions and discussions (Harrison, 2014). While this suggests more possibilities for evidence collection, it also can be problematic as teachers juggle the collection, meaning and importance of the data they are able to harvest.

Formative assessment provides opportunities for teachers to elicit anticipated and emergent information about student understanding and where they can make judgements and use assessment evidence to create feedback interactions through classroom routines (Cowie et al., 2018). Teachers interpret and make decisions about such assessment evidence in a timely fashion that can drive planning and support next stages of learning (Black et al., 2003; Correia & Harrison, 2020). Within formative assessment cycles, feedback constitutes the link between this assessment and the adaptation of teaching (Bernholt et al., 2012). Sadler (1989) defines feedback as information about the gap between the actual performance level and the reference level. Feedback indicates to the learner how successful something has been or how successfully it has been done. It may be initiated from specific events or through interactions with students within the inquiry activity, enabling the teacher to assess new evidence in relation to earlier pieces (Shute, 2008). This dynamic process opens up a way of deciding on next steps using recent evidence set within the broader range of evidence that a teacher has built up for particular children and classes (Hattie & Timperley, 2007). In this way, as new assessment evidence is collected, teachers are able to map that information against previous evidence they hold about an individual's previous performance in that area and against anticipated range of proficiency they have envisaged, from experiencing other learners completing that activity.

Teacher knowledge plays a key role when assessment is used to inform, shape and monitor teaching practices. Teachers need to recognise achievement within a particular inquiry context, while also being cognisant with what progress might look like for the specific inquiry competences involved (Black et al., 2011). In this sense, assessment is no longer a separate activity from teaching and learning but, instead, is dynamically embedded as part of classroom activity since inquiry-learning and formative assessment are coordinated (Ruiz-Primo, 2011). Within favourable conditions, the inquiry context provides the conditions for eliciting evidence of current understanding and opportunity for judgements for both teachers and learners; this allows for responsive action, in the time frame of the activity or lesson (Furtak & Heredia, 2014). While blending inquiry and formative assessment approaches seems compatible, asking teachers to change both aspects together challenges their regular teaching routines and practices as well as their professional knowledge and beliefs.

Importance of informal formative assessment

A formative approach to assessment encourages greater student agency because it provides classroom activities that generate, reveal and value student thinking and allows students to be involved in decisions about learning. The implication for teachers is that they need to be sensitive to opportunities for students to exercise conceptual agency and support students in developing skills and confidence to recognise how their own thinking is playing out in relation to ideas raised by others in the classroom (Cowie et al., 2013, 2018). It is therefore a complex and demanding process, one in which both teachers and students need to be active and informed participants and one in which attention to student activity through observation and interaction is vital.

Conversations with students, as they work on an inquiry task, provide teachers with a rich bed of evidence of students' current and emerging thinking (Coffey et al., 2011; Cowie & Bell, 1999). These practices encourage students to articulate and share their ideas as they unfold during classroom activities, recognising emerging ideas and encouraging individuals and peers to share their responses, so co-constructing knowledge and thus providing feedback to move students forward in their learning (Zuiker & Whitaker, 2014). This act of eliciting evidence followed by noticing – recognising and responding to students' ideas during instruction – is the driver for formative assessment, where teachers and students use assessment evidence to tailor future teaching and learning (Cowie et al., 2018; Mason, 2002; Ruiz-Primo & Furtak, 2007).

Reference points for specific pieces of assessment evidence help teachers notice particular features of student performance; this process enables them to make nuanced judgements about what they consider would be productive next steps in learning for individuals, groups and the whole class (Cowie et al., 2018). At the same time, students are also able to reflect on their current and past understanding, competences and skills, within a topic, and realise where they are improving and where they still require support with their learning (Andrade, 2010). The key point here, is that the evidence is elicited and considered through feedback loops generated through the many levels of interaction that happen in the classroom, as learning is taking place.

In this special issue, Nieminen et al. (2020) explore how this kind of informal formative assessment based on classroom discussions or on interactions between teachers and students entails diverse practices according both to the forms and the functions dedicated to these episodes by teachers, to what teachers notice and value and by their pedagogical decisions in the moment. Rached and Grangeat (2020) analyse how informal formative assessment is also influenced by the way the students engage in the discussion and note the variation in practice among teachers.

Supporting students in peer- and self-assessment

Inquiry activities both generate and sustain a focus on student thinking through the interactions that occur throughout the inquiry process. Through discussing with others, a learner begins to get a view of how much and how well they understand the ideas within a topic. These interactions challenge existing ideas and also shape and adapt existing knowledge as the student makes meaning with others, with peers providing both a resource and support for learning (Arnold et al., 2014).

By developing their peer and self-assessment skills, formative assessment supports students to be more active in the assessment process (Tsivitanidou et al., 2018). This fosters students' engagement, commitment, and effort because they need to instigate the process where one another's contributions can be voiced, heard, and reviewed. In most STEM classrooms, formalised peer and self-assessment exercises need to be modelled for students to help them understand their roles and the role of their peers in these situations (Correia & Harrison, 2020). Such practices also require teachers to be aware of what evidence to look for and next steps in the progression of learning. This often requires them using specific criteria and reference points to make these assessments and provide appropriate feedback.

Peer and self-assessment build on how teachers provide feedback to individuals by offering a frame and model that influences the phrases and terminology used. These practices set the style of language and the tone for providing feedback and students tend to adopt some of these approaches when engaging in peer feedback activities. Part of this arises naturally within the STEM classroom climate (Peters, 2013) as students note the specific ways their teacher provides feedback both orally and in written form; they can learn from these events and mimic the tone, language and ways in which the teacher interacts with students in class (Shute, 2008). In this special issue, Grob et al. (2019) examine the extent to which these methods should be adapted to the particular school level and context. They determine how the implementation of formative assessment in a primary school, where science is taught as an integrated subject by a single teacher, differs from that of a secondary school with one teacher per scientific discipline, sometimes with a very small weekly timetable.

Implementing new teaching approaches

Adapting teaching practices that integrate assessment within an inquiry approach is far from being a simple task. Simon et al. (2006) indicate that teachers face difficulties when adopting new teaching approaches which conflict with existing beliefs about teaching and learning. Teachers need to consider how to change practice in ways that create opportunities for students to also collect, interpret and re-consider evidence. In inquiry classrooms, teacher-student interactions need to both probe and instigate thinking and reflection. Developing appropriate teaching skills to sustain this interplay between current ideas and other possibilities, in a manner that remains true to the inquiry spirit, is a challenge for many teachers and their students. Making this shift in thinking to the idea that assessment opportunities are embedded within instruction can bring about a paradigm shift in what takes place in the classroom. This can affect the types of activity undertaken and how those activities are organised and implemented. For example, teachers might wish to increase or adapt the way they do group work, in order for students to use their peers as a resource that provides and checks ideas as they arise within classroom activities (Heritage, 2007).

There are a number of different approaches that distinguish how formative practices enable teachers and learners to engage with an understanding of inquiry competence and that encourage learners to improve. One essential feature is that the inquiry task is planned and facilitated in order to promote dialogue that will help elicit evidence of student understanding, rather than assuming that a particular level of understanding is shared across all the students in the class (Black & Wiliam, 2018). In addition, to make explicit their learning goals and to monitor students' progress in learning, teachers often use a rubric that outlines the expected competences and indicates levels of attainment described in a qualitative sense (Andrade et al., 2008). Such rubrics can provide teachers with perspectives on the development of competences and guide students to the next level through feedback. While this can be useful in enabling teachers to visualise likely trajectories within specific inquiry activities, it can also be problematic if feedforward is constrained by too strict a goal orientation, which gives no room for student deviations and creativity. In this special issue, Ropohl and Rönnebeck (2019) explore the way pre-service teachers interpret students' ideas and propose helpful comments that may underpin shifts in reasoning. Other teachers entrust students with the responsibility of assessing their peers' work. The teacher has to make decisions about which pieces of evidence to act on and judge which ideas are contingent for future learning both for individuals and groups of students. Depending on the situation, one or more of these ideas might be taken forward or they might be parked for a future lesson and the evidence then possibly used to connect activities as well as foster genuine deep learning.

What is challenging in this approach to assessment is the collection and verification of the assessment evidence, since much of it is ephemeral, and evidence is continually being updated as the inquiry proceeds (Kawalkar & Vijapurkar, 2013). The problem for the teacher is ascertaining whether students realise that they have used a better approach and how alternative methods relate to one another. Ascertaining the reasons for a solution or a decision is more difficult than simply checking if students can give a correct response to a question in a test. Therefore, the style of assessment that accompanies inquiry, is often a complex and intricate process and set of skills for the teachers to develop and use effectively (Zuiker & Whitaker, 2014). This becomes even more difficult when students are content to reach a solution and rather than question, justify and articulate why that a particular approach is better. Osborne (2011, p. 100) in his study on using argumentation to explore evidence with students reported that:

> The task confronting the teacher is not just one of convincing the student of why the scientific account is correct but also one of convincing them of why alternative theoretical accounts, including their own misconceptions, are wrong.

Effective formative assessment, therefore, depends on understanding of processes involved within a particular inquiry context, a recognition of those competencies capable of being developed and a notion of progression in the competences that form the learning goals and that lead to their acquisition. Making explicit and visible the learning goals and the progression of competence in inquiry is a complex task that requires teachers to be aware of how students develop their learning. Currently, there is little in the literature that helps teachers understand and recognise the development of inquiry competence (Schneider & Plasman, 2011). However, from their own knowledge acquisition and the practice-based evidence that they have witnessed with their students, teachers recognise some of the stages on an inquiry learning progression for specific inquiry activities that they are familiar with (Harrison, 2014). With this in mind, teachers might be able to construct or use assessment tools that have the potential to situate students along a progression, albeit rudimentary.

For such actions to occur, teachers need to plan carefully how they set up and organise lessons for inquiry learning. Minner et al. (2010) showed that student learning is more likely to occur when they can engage in hands-on activities with science phenomena and are entrusted with responsibility for their own learning. Such practices led to higher order thinking for students. Nevertheless, such opportunities do not automatically occur when students are asked to steer an inquiry, resolve a problem, or realise a project. Often students require considerable teacher support and guidance in the early stages of inquiry learning as they begin to try out new ideas and new ways of working.

The difficulty for the teacher is determining which approach to take, particularly when they are already making quite radical changes in their teaching through adopting an inquiry approach. Some guidelines and resources exist on how to incorporate assessment within inquiry-based teaching explored by science education research projects (Correia & Harrison, 2020; Harrison, 2014). Nevertheless, these practices are innovative and only just emerging as accepted practices.

Implementing assessment in inquiry-learning

Combining inquiry-learning with formative and summative assessment results in a complex teaching approach. It differs markedly from previous instructional approaches by requiring a different specific pattern of interaction to be developed to encourage students to reveal their developing thinking as they engage in the inquiry process. This results in a shift both in teaching practices, knowledge about the content and the way it is learn, and also in teachers' beliefs about science education (Furtak & Alonzo, 2010). Some teachers find it difficult to conceptualise the classroom routines and climate that inquiry and formative assessment require to function effectively, and this can lead to a reluctance to introduce new ideas. Moreover, the alignment of formative and summative assessments is not easy to achieve and implementing them in a complementary manner remains a challenge for teachers and researchers (Dixson & Worrell, 2016). Teachers need ongoing professional support, time, and confidence, in order to reflect critically upon and to develop their practice since their approaches and skills of teaching become more refined over time (Schneider & Plasman, 2011).

The literature on teacher development suggests that change is a slow and difficult process, and even more so when assessment practices are involved (Hargreaves & Fullan, 2012; Harrison, 2013). This makes implementing a new pedagogical approach combining inquiry-based learning with new assessment methods rather difficult. It is not simply a matter of adding on to existing practice – it requires a reconsideration of practice to create and enhance the necessary interventions to improve learning and assessment. Three factors emerge from the studies as being favourable for maximising the effect of teacher professional development (Lotter et al., 2016). First, the shift of teachers' practices needs to be supported as it develops, to allow teachers to gradually explore new ideas and trial new ways of presenting relevant content and new teaching and assessment methods (Olin & Ingerman, 2016; Schleicher, 2015). Therefore, teacher development programmes have to be designed as an on-going process that lasts at least a semester to allow for such developments. Second, while time is necessary, it is not sufficient, and the shift of practices is also supported by professional social interactions. Teacher collaboration and co-operation with teacher educators and researchers is beneficial for teachers to express their difficulties, be supported and share their efforts to improve their practices (Akerson et al., 2009; Grangeat & Gray, 2008; Harrison et al., 2008; Lepareur & Grangeat, 2018; Mottier Lopez & Pasquini, 2017). Third, the focus of the programme is a crucial key, keeping a gentle balance between updating teacher content knowledge and promoting generic pedagogic strategies. Focusing both on specific presentations of content and design of teaching methods involves teachers in a dynamic of change that fits better with their actual professional concerns than teacher development programmes that focus only on one of these aspects (Grangeat, 2015; Nilsson, 2014; Van Driel & Berry, 2012).

Implementing collaboration between teachers and researchers

The articles in this special issue report on collaborative teacher-researcher research designed to fashion new ways of learning content through inquiry and relevant pedagogical methodology through a more formative approach to assessment. Throughout the ASSIST-ME project, teachers had been involved working with research teams for extended periods, in order to collectively design, trial and elaborate teaching plans that were trialled and progressively refined. Local Working Groups (LWGs), consisting of researchers and teachers, developed assessment and analytical tools to support research of specific assessment methods in inquiry classrooms; these LWGs existed and met regularly for two school years. While each category has their role to play in this relationship, teachers and researchers work together to define and carry out research that both find beneficial. Their interest lay in how useful each method was in collecting data on student understanding and achievement, how tools were implemented in the classroom and how effective each method was in providing feedback for teachers and students. This organisation of the research process based on local working groups and inclusive participatory strategies is cohesive with the European Responsible Research and Innovation approach since the public and the recipients of research outcomes are engaged in the design of some of the research tools and methods (Hazelkorn et al., 2015). Conversely, it allows for changes in assessment culture that could be adapted flexibly for specific contexts (Dolin et al., 2018).

In this issue, Zlabkova et al. (2020) explore how teachers' perspective on formative assessment and their readiness to implement the new forms of assessment in their teaching evolves. The key to success lies in tailoring and adjusting professional learning so that sufficient and varied exemplars of inquiry teaching and assessment methods can be explored, and evidence of emerging practice shared and supported with other practitioners. The collective reflections and mutual support were of great importance for the teachers to carry on and persist.

Conclusion

This special issue intends to deepen understanding of how formative assessment methods may adapt different features within STEM classroom practices and how formative and summative assessment could be coordinated. While previous research has noted that teachers generally accept that a formative approach to classroom assessment is beneficial, they do not always manage to restructure the way they teach to accommodate formative practice (Swaffield, 2011). Being aware of and researching the challenges that face teachers when they attempt to meld inquiry pedagogy with formative assessment is important. Moreover, evaluating how such changes develop and work in the classroom is a key part of implementing such practices within the teaching profession. Therefore, the articles explore three domains:

- Informal formative assessment that may occur in any student-teacher or student-student interaction is mainly addressed through the lens of classroom conversations. How these interactions are set up, managed and then used as evidence for formative practice is considered from two different perspectives (a) by investigating variations in the practices of two teachers (Nieminen et al., 2020); (b) by exploring teachers' and students' insights in classroom conversations (Rached & Grangeat, 2020).
- Structured formative assessment is addressed through lessons when teachers or peers collect information from students at a planned time and then later analyse the collected information to provide feedback. Two formative assessment methods are analysed: (a) teacher written feedback (Ropohl & Rönnebeck, 2019); (b) teaching plans and materials, at primary and secondary schools (Grob et al., 2019).
- Ongoing evolution of teacher knowledge and practices of teachers who had been co-investigating their own practice (in research project) is evaluated to better understand to what extent they succeed in integrating assessment into their pedagogic practices (Zlabkova et al., 2020).

This special issue complements the findings of the ASSIST-ME European research project (Dolin & Evans, 2018). The articles illustrate how ideas about inquiry and assessment are played out in the classroom, revealing both how the pedagogical approach fits with the assessment choices made and the support that teachers needed, in order to engage with these changes. It both informs and presents some of the challenges related to the uptake and development of inquiry learning in STEM classrooms. It also raises the question of how national educational systems can foster and build the systemic support measures and tools that teachers need to integrate assessment into their classroom practices that fit alongside or within their summative assessment practices.

Disclosure statement

No potential conflict of interest was reported by the author(s).

Funding

This work was supported by FP7 [grant number 321428].

ORCID

Michel Grangeat http://orcid.org/0000-0002-1951-6547
Christine Harrison http://orcid.org/0000-0002-1675-6107

References

Akerson, V. L., Cullen, T. A., & Hanson, D. L. (2009). Fostering a community of practice through a professional development program to improve elementary teachers' views of nature of science and teaching practice. *Journal of Research in Science Teaching*, *46*(10), 1090–1113. https://doi.org/10.1002/tea.20303

Andrade, H. L. (2010). Students as the definitive source of formative assessment: Academic self-assessment and the self-regulation of learning. *NERA Conference Proceedings*, *25*, 19. https://opencommons.uconn.edu/nera_2010/25

Andrade, H. L., Du, Y., & Wang, X. (2008). Putting rubrics to the test: The effect of a model, criteria generation, and rubric-referenced self-assessment on elementary school students' writing. *Educational Measurement: Issues and Practice*, *27*(2), 3–13. https://doi.org/10.1111/j.1745-3992.2008.00118.x

ARG. (2002). *Assessment for learning: 10 principles*. University of Cambridge.

Arnold, J. C., Kremer, K., & Mayer, J. (2014). Understanding students' experiments—What kind of support do they need in inquiry tasks? *International Journal of Science Education*, *36*(16), 2719–2749. https://doi.org/10.1080/09500693.2014.930209

Artigue, M., & Blomhøj, M. (2013). Conceptualizing inquiry-based education in mathematics. *ZDM Mathematics Education*, *45*(6), 797–810. https://doi.org/10.1007/s11858-013-0506-6

Barron, B., & Darling-Hammond, L. (2008). *Teaching for meaningful learning: A review of research on inquiry-based and cooperative learning. Book excerpt*. George Lucas Educational Foundation.

Bell, B., & Cowie, B. (2001). The characteristics of formative assessment in science education. *Science Education*, *85*(5), 536–553. https://doi.org/10.1002/sce.1022

Bernholt, S., Neumann, K., & Nentwig, P. (Eds.). (2012). *Making it tangible learning outcomes in science education*. Waxmann.

Biggers, M., & Forbes, C. T. (2012). Balancing teacher and student roles in elementary classrooms: Preservice elementary teachers' learning about the inquiry continuum. *International Journal of Science Education*, *34*(14), 2205–2229. https://doi.org/10.1080/09500693.2012.694146

Black, P., Harrison, C., Clare, L., Bethan, M., & Dylan, W. (Eds.). (2003). *Assessment for learning: Putting it into practice*. Open University Press.

Black, P., & Wiliam, D. (2018). Classroom assessment and pedagogy. *Assessment in Education: Principles, Policy & Practice*, *25*(6), 551–575. https://doi.org/10.1080/0969594X.2018.1441807

Black, P., Wilson, M., & Yao, S.-Y. (2011). Road maps for learning: A guide to the navigation of learning progressions. *Measurement: Interdisciplinary Research and Perspectives*, *9*(2–3), 71–123. https://doi.org/10.1080/15366367.2011.591654

Chin, C. (2006). Classroom interaction in science: Teacher questioning and feedback to students' responses. *International Journal of Science Education*, *28*(11), 1315–1346. https://doi.org/10.1080/09500690600621100

Coffey, J. E., Hammer, D., Levin, D. M., & Grant, T. (2011). The missing disciplinary substance of formative assessment. *Journal of Research in Science Teaching*, *48*(10), 1109–1136. https://doi.org/10.1002/tea.20440

Correia, C. F., & Harrison, C. (2020). Teachers' beliefs about inquiry-based learning and its impact on formative assessment practice. *Research in Science and Technological Education*, *38*(3), 355–376. https://doi.org/10.1080/02635143.2019.1634040

Cowie, B., & Bell, B. (1999). A model of formative assessment in science education. *Assessment in Education: Principles, Policy & Practice*, *6*(1), 101–116. https://doi.org/10.1080/09695949993026

Cowie, B., Harrison, C., & Willis, J. (2018). Supporting teacher responsiveness in assessment for learning through disciplined noticing. *The Curriculum Journal*, *29*(4), 464–478. https://doi.org/10.1080/09585176.2018.1481442

Cowie, B., Moreland, J., & Otrel-Cass, K. (2013). *Expanding notions of assessment for learning: Inside science and technology primary classrooms*. Springer Science & Business Media.

Dixson, D. D., & Worrell, F. C. (2016). Formative and summative assessment in the classroom. *Theory Into Practice*, *55*(2), 153–159. https://doi.org/10.1080/00405841.2016.1148989

Dolin, J., Black, P., Harlen, W., & Tiberghien, A. (2018). Exploring relations between formative and summative assessment. In J. Dolin & R. H. Evans, *Transforming assessment* (pp. 53–80). Springer.

Dolin, J., & Evans, R. (Eds.). (2018). *Transforming assessment: Through an interplay between practice, research and policy*. Springer International Publishing. https://doi.org/10.1007/978-3-319-63248-3

Dolin, J., & Krogh, L. B. (2010). The relevance and consequences of PISA science in a Danish context. *International Journal of Science and Mathematics Education*, *8*(3), 565–592. https://doi.org/10.1007/s10763-010-9207-6

Furtak, E. M., & Alonzo, A. C. (2010). The role of content in inquiry-based elementary science lessons: An analysis of teacher beliefs and enactment. *Research in Science Education*, *40*(3), 425–449. https://doi.org/10.1007/s11165-009-9128-y

Furtak, E. M., & Heredia, S. C. (2014). Exploring the influence of learning progressions in two teacher communities. *Journal of Research in Science Teaching*, *51*(8), 982–1020. https://doi.org/10.1002/tea.21156

Furtak, E. M., Kiemer, K., Circi, R. K., Swanson, R., de León, V., Morrison, D., & Heredia, S. C. (2016). Teachers' formative assessment abilities and their relationship to student learning: Findings from a four-year intervention study. *Instructional Science*, *44*(3), 267–291. https://doi.org/10.1007/s11251-016-9371-3

Grangeat, M. (2015). Exploring the set of pedagogical knowledge, from pedagogy to content. In M. Grangeat (Ed.), *Understanding science teachers' professional knowledge growth* (pp. 117–133). Sense Publishers.

Grangeat, M. (2016). Dimensions and modalities of inquiry-based teaching: Understanding the variety of practices. *Education Inquiry*, *7*(4), 421–442. https://doi.org/10.3402/edui.v7.29863

Grangeat, M., & Gray, P. (2007). Factors influencing teachers' professional competence development. *Journal of Vocational Education & Training*, *59*(4), 485–501. https://doi.org/10.1080/13636820701650943

Grangeat, M., & Gray, P. (2008). Teaching as a collective work: Analysis, current research and implications for teacher education. *Journal of Education for Teaching*, *34*(3), 177–189. https://doi.org/10.1080/02607470802212306

Greene, M., & von Renesse, C. (2017). A path to designing inquiry activities in mathematics. *PRIMUS*, *27*(7), 646–668. https://doi.org/10.1080/10511970.2016.1211203

Grob, R., Holmeier, M., & Labudde, P. (2019). Analysing formal formative assessment activities in the context of inquiry at primary and upper secondary school in Switzerland. *International Journal of Science Education*, *43*(3), 407–427. https://doi.org/10.1080/09500693.2019.1663453

Hargreaves, A., & Fullan, M. (2012). *Professional capital: Transforming teaching in every school.* Teachers College Press.

Harlen, W. (2009). Teaching and learning science for a better future. *School Science Review*, *90* (333), 33–41.

Harrison, C. (2013). Changing assessment practices in science classrooms. In D. Corrigan, R. Gunstone, & A. Jones (Eds.), *Valuing assessment in science education: Pedagogy, curriculum, policy* (pp. 347–358). Springer Netherlands. https://doi.org/10.1007/978-94-007-6668-6_18

Harrison, C. (2014). Assessment of inquiry skills in the SAILS project. *Science Education International*, *25*(1), 112–122.

Harrison, C., Hofstein, A., Eylon, B.-S., & Simon, S. (2008). Evidence-based professional development of Science teachers in two countries. *International Journal of Science Education*, *30*(5), 577–591. https://doi.org/10.1080/09500690701854832

Hattie, J., & Timperley, H. (2007). The power of feedback. *Review of Educational Research*, *77*(1), 81–112. https://doi.org/10.3102/003465430298487

Hazelkorn, E., Ryan, C., Beernaert, Y., Constantinou, C., Deca, L., Grangeat, M., Karikorpi, M., Lazoudis, A., Pinto-Casulleras, R., & Welzel-Breuer, M. (2015). *Science education for responsible citizenship* (EUR 26893). European Commission - Research and Innovation.

Heinz, J., Enghag, M., Stuchlikova, I., Cakmakci, G., Peleg, R., & Baram-Tsabari, A. (2017). Impact of initiatives to implement science inquiry: A comparative study of the Turkish, Israeli, Swedish and Czech science education systems. *Cultural Studies of Science Education*, *12*(3), 677–708. https://doi.org/10.1007/s11422-015-9704-6

Heritage, M. (2007). Formative assessment: What Do teachers need to know and do? *Phi Delta Kappan*, *89*(2), 140–145. https://doi.org/10.1177/003172170708900210

Kawalkar, A., & Vijapurkar, J. (2013). Scaffolding science talk: The role of teachers' questions in the inquiry classroom. *International Journal of Science Education*, *35*(12), 2004–2027. https://doi.org/10.1080/09500693.2011.604684

Krajcik, J., McNeill, K. L., & Reiser, B. J. (2008). Learning-goals-driven design model: Developing curriculum materials that align with national standards and incorporate project-based pedagogy. *Science Education, 92*(1), 1–32. https://doi.org/10.1002/sce.20240

Lepareur, C., & Grangeat, M. (2018). Teacher collaboration's influence on inquiry-based science teaching methods. *Education Inquiry, 9*(4), 363–379. https://doi.org/10.1080/20004508.2018.1428037

Lin, K.-Y., Hsiao, H.-S., Williams, P. J., & Chen, Y.-H. (2020). Effects of 6E-oriented STEM practical activities in cultivating middle school students' attitudes toward technology and technological inquiry ability. *Research in Science & Technological Education, 38*(1), 1–18. https://doi.org/10.1080/02635143.2018.1561432

Linn, M. C., Davis, E. A., & Bell, P. (2004). *Internet environments for science education*. Lawrence Erlbaum.

Looney, J. W. (2011). *Integrating formative and summative assessment: Progress toward a seamless system?* (No. 58; OECD Education Working Papers). OECD. http://doi.org/10.1787/5kghx3kbl734-en

Lotter, C., Smiley, W., Thompson, S., & Dickenson, T. (2016). The impact of a professional development model on middle school science teachers' efficacy and implementation of inquiry. *International Journal of Science Education, 38*(18), 2712–2741. https://doi.org/10.1080/09500693.2016.1259535

Mason, J. (2002). *Researching your own practice: The discipline of noticing*. Routledge.

Matthews, P. S. C., & McKenna, P. J. (2005). Assessment of practical work in Ireland: A critique. *International Journal of Science Education, 27*(10), 1211–1224. https://doi.org/10.1080/09500690500102573

Minner, D. D., Levy, A. J., & Century, J. (2010). Inquiry-based science instruction–what is it and does it matter? Results from a research synthesis years 1984 to 2002. *Journal of Research in Science Teaching, 47*(4), 474–496. https://doi.org/10.1002/tea.20347

Mottier Lopez, L., & Pasquini, R. (2017). Professional controversies between teachers about their summative assessment practices: A tool for building assessment capacity. *Assessment in Education: Principles, Policy & Practice, 24*(2), 228–249. https://doi.org/10.1080/0969594X.2017.1293001

Nieminen, P., Hähkiöniemi, M., & Viiri, J. (2020). Forms and functions of on-the-fly formative assessment conversations in physics inquiry lessons. *International Journal of Science Education, 43*(3), 362–384. https://doi.org/10.1080/09500693.2020.1713417

Nilsson, P. (2014). When teaching makes a difference: Developing science teachers' pedagogical content knowledge through learning study. *International Journal of Science Education, 36*(11), 1794–1814. https://doi.org/10.1080/09500693.2013.879621

Olin, A., & Ingerman, Å. (2016). Features of an emerging practice and professional development in a science teacher team collaboration with a researcher team. *Journal of Science Teacher Education, 27*(6), 607–624. https://doi.org/10.1007/s10972-016-9477-0

Osborne, J. (2011). Science teaching methods: A rationale for practices. *School Science Review, 93*(343), 93–103.

Pellegrino, J. W., Chudowsky, N., & Glaser, R. (2001). *Knowing what students know: The science and design of educational assessment*. National Academy Press. https://doi.org/10.17226/10019

Peters, M. L. (2013). Examining the relationships among classroom climate, self-efficacy and achievement in undergraduate mathematics: A multi-level analysis. *International Journal of Science and Mathematics Education, 11*(2), 459–480. https://doi.org/10.1007/s10763-012-9347-y

Rached, E., & Grangeat, M. (2020). French teachers' informal formative assessment in the context of inquiry-based learning. *International Journal of Science Education, 43*(3), 385–406. https://doi.org/10.1080/09500693.2020.1740818

Rocard, M., Cesrmley, P., Jorde, D., Lenzen, D., & Hemmo, V. (2007). *Science education now: A renewed pedagogy for the future of Europe*. European Commission.

Ropohl, M., & Rönnebeck, S. (2019). Making learning effective – Quantity and quality of pre-service teachers' feedback. *International Journal of Science Education, 41*(15), 2156–2176. https://doi.org/10.1080/09500693.2019.1663452

Rönnebeck, S., Bernholt, S., & Ropohl, M. (2016). Searching for a common ground – A literature review of empirical research on scientific inquiry activities. *Studies in Science Education*, *52*(0), 161–197. https://doi.org/10.1080/03057267.2016.1206351

Ruiz-Primo, M. A. (2011). Informal formative assessment: The role of instructional dialogues in assessing students' learning. *Studies in Educational Evaluation*, *37*(1), 15–24. https://doi.org/10.1016/j.stueduc.2011.04.003

Ruiz-Primo, M. A., & Furtak, E. M. (2007). Exploring teachers' informal formative assessment practices and students' understanding in the context of scientific inquiry. *Journal of Research in Science Teaching*, *44*(1), 57–84. https://doi.org/10.1002/tea.20163

Sadler, D. R. (1989). Formative assessment and the design of instructional systems. *Instructional Science*, *18*(2), 119–144. https://doi.org/10.1007/BF00117714

Schleicher, A. (2015). *Schools for 21st-century learners*. https://doi.org/10.1787/9789264231191-en

Schneider, R. M., & Plasman, K. (2011). Science teacher learning progressions: A review of science teachers' pedagogical content knowledge development. *Review of Educational Research*, *81*(4), 530–565. https://doi.org/10.3102/0034654311423382

Shavelson, R. J., Young, D. B., Ayala, C. C., Brandon, P. R., Furtak, E. M., Ruiz-Primo, M. A., Tomita, M. K., & Yin, Y. (2008). On the impact of curriculum-embedded formative assessment on learning: A collaboration between curriculum and assessment developers. *Applied Measurement in Education*, *21*(4), 295–314. https://doi.org/10.1080/08957340802347647

Shute, V. J. (2008). Focus on formative feedback. *Review of Educational Research*, *78*(1), 153–189. https://doi.org/10.3102/0034654307313795

Simon, S., Erduran, S., & Osborne, J. (2006). Learning to teach argumentation: Research and development in the science classroom. *International Journal of Science Education*, *28*(2–3), 235–260. https://doi.org/10.1080/09500690500336957

Stewart, J., Cartier, J. L., & Passmore, C. M. (2005). Developing understanding through model-based inquiry. In M. S. Donovan & J. D. Bransford (Eds.), *How students learn: Science in the classroom* (pp. 515–565). National Academy Press. https://doi.org/10.17226/11102

Swaffield, S. (2011). Getting to the heart of authentic assessment for learning. *Assessment in Education: Principles, Policy & Practice*, *18*(4), 433–449. https://doi.org/10.1080/0969594X.2011.582838

Tsivitanidou, O. E., Constantinou, C. P., Labudde, P., Rönnebeck, S., & Ropohl, M. (2018). Reciprocal peer assessment as a learning tool for secondary school students in modelling-based learning. *European Journal of Psychology of Education*, *33*(1), 51–73. https://doi.org/10.1007/s10212-017-0341-1

Van Driel, J. H., & Berry, A. (2012). Teacher professional development focusing on pedagogical content knowledge. *Educational Researcher*, *41*(1), 26–28. https://doi.org/10.3102/0013189X11431010

Waddington, D., Nentwig, P., & Schanze, S. (2007). *Making it comparable: Standards in science education*. Waxmann Verlag GmbH.

Wiliam, D. (2011). What is assessment for learning? *Studies in Educational Evaluation*, *37*(1), 3–14. https://doi.org/10.1016/j.stueduc.2011.03.001

Windschitl, M., Thompson, J., & Braaten, M. (2008). Beyond the scientific method: Model-based inquiry as a new paradigm of preference for school science investigations. *Science Education*, *92*(5), 941–967. https://doi.org/10.1002/sce.20259

Zeng, W., Huang, F., Yu, L., & Chen, S. (2018). Towards a learning-oriented assessment to improve students' learning—A critical review of literature. *Educational Assessment, Evaluation and Accountability*, *30*(3), 211–250. https://doi.org/10.1007/s11092-018-9281-9

Zlabkova, I., Petr, J., Stuchlikova, I., Rokos, L., & Hospesova, A. (2020). Development of teachers' perspective on formative peer assessment. *International Journal of Science Education*, *43*(3), 428–448. https://doi.org/10.1080/09500693.2020.1713418

Zuiker, S., & Whitaker, J. R. (2014). Refining inquiry with multi-form assessment: Formative and summative assessment functions for flexible inquiry. *International Journal of Science Education*, *36*(6), 1037–1059. https://doi.org/10.1080/09500693.2013.834489

Forms and functions of on-the-fly formative assessment conversations in physics inquiry lessons

Pasi Nieminen, Markus Hähkiöniemi and Jouni Viiri

ABSTRACT

This case study examined teachers' on-the-fly formative assessment conversations, that is, how teachers collect information from students' thinking and use that information to support their learning during teacher-student interactions. Previous studies have typically analysed whole-class discussions on the level of speaking turns, and have identified means of collecting information (eliciting) and using information (feedback). This study used episode-based analysis to identify both forms and functions of on-the-fly formative assessment conversations during all phases of the inquiry lessons. We video-recorded two teachers' physics lessons for 7th graders (13–14 years). When analysing the forms of on-the-fly formative assessment conversations, we found five types of collecting information, including questioning and other means, and four ways of using the information. We also studied the functions of on-the-fly formative assessment and found four main functions and their sub-functions. The main functions differed according to the purpose of guiding in relation to the flow of student inquiry. Teachers (1) checked or examined students' thinking, (2) corrected or developed students' present thinking, (3) redirected students' thinking and (4) reviewed students' thinking. The analysis of forms and functions enabled us to recognise differences in two physics inquiry lessons on the same topic.

Introduction

Assessment of student learning can be performed by diverse formative and summative practices depending on the pedagogical purposes (Wiliam & Black, 1996; Brookhart, 2015; Dolin, Black, Harlen, & Tiberghien, 2018) and learning goals of assessment, such as subject-specific and generic competences (Rönnebeck, Nielsen, Olley, Ropohl, & Stables, 2018). Assessment of inquiry-based learning, which has been advocated in educational research and policy over the past decade (Rocard et al., 2007), requires methods that meet the competences and learning goals of inquiry. One promising practice to assess students' learning in their daily work is on-the-fly formative assessment, which takes place in teacher–student conversations (Ruiz-Primo, 2011; Shavelson et al., 2008). Previous studies of on-the-fly formative assessment have typically focused on whole-class discussions (e.g. Ruiz-Primo & Furtak, 2006; Furtak, Ruiz-Primo, & Bakeman,

2017). The purpose of our qualitative case study was to explore and understand two teachers' on-the-fly formative assessment practices when they were supporting student groups' inquiry process during a task for image formation in a plane mirror.

On-the-fly formative assessment conversations

Assessment in classrooms is usually divided into summative and formative assessment according to their purposes (Wiliam & Black, 1996; Brookhart, 2015; Dolin et al., 2018). Summative assessment is 'assessment *of* learning' as its purpose is to summarise, value and report what has been learnt after a learning period. Formative assessment, in contrast, is 'assessment *for* learning' as it attempts to enhance and support student learning during the learning process. Formative and summative assessment can be seen as the extremes of a continuum since, in some cases, the same data can be used formatively and summatively (Wiliam & Black, 1996; Dolin et al., 2018).

There are two frequently cited definitions for formative assessment that include common characteristics. Both of the following citations emphasise the process through which teachers and students attempt to understand the current state of learning and how to achieve the goal.

> Assessment for Learning is the process of seeking and interpreting evidence for use by learners and their teachers to decide where the learners are in their learning, where they need to go and how best to get there. (Assessment Reform Group, 2002, p. 2)

> Practice in a classroom is formative to the extent that evidence about student achievement is elicited, interpreted, and used by teachers, learners, or their peers, to make decisions about the next steps in instruction that are likely to be better, or better founded, than the decisions they would have taken in the absence of the evidence that was elicited. (Black & Wiliam, 2009, p. 9)

Further, formative assessment practices can be considered to lie on a continuum according to the degree of *formality*. Formality refers to the amount of planning and the manner in which the information is collected and the feedback is given (Bell & Cowie, 2001; Shavelson et al., 2008; Ruiz-Primo, 2011; Ruiz-Primo & Kroog, 2017). An example of *formal* formative assessment is a teacher's pre-planned test, which students answer during a lesson to inform themselves and the teacher about the current understanding without summative records. *Informal* formal assessment is, for example, when the teacher elicits students' understanding via conversation or asks them to raise hands and uses that information to help them to proceed.

Informal formative assessment has been referred to by the terms interactive formative assessment (Bell & Cowie, 2001), assessment conversations (Ruiz-Primo & Furtak, 2006) and on-the-fly formative assessment (Shavelson et al., 2008). In this paper, we use the term *on-the-fly assessment* or, briefly, *on-the-fly*. All these terms refer to the interaction between a teacher and students, which is not wholly planned beforehand, but takes place in real-time when the teacher recognises appropriate opportunities to support students in advancing their learning. According to Shavelson et al. (2008), 'on-the-fly formative assessment arises when a "teachable moment" unexpectedly occurs' (p. 300). To some extent, teachers can prepare in advance and increase opportunities to interact with students but 'they do not know the details of when and what assessment will occur' (Bell & Cowie, 2001, p. 541). If teachers have sufficient content knowledge and pedagogical content knowledge to react in real time to student contributions, on-the-fly assessment can be effective for learning (Ruiz-Primo, 2011).

There are some studies directly addressing the concept of on-the-fly, but there is a larger body of research on teacher–student interaction, which is an essential part of on-the-fly. Often, classroom discussions tend to follow a certain pattern called IRE: teacher initiation, student response and teacher evaluation (Mehan, 1979). It has been stated that this pattern is related to the teacher's authoritative manner when looking for a certain answer that she wants to hear from students (Mortimer & Scott, 2003). Instead of evaluation (E), the teacher may offer elaborative feedback or pose a follow-up move (F) for the student's response, which produces an I-R-F pattern or an even longer chain, I-R-F-R-F-. Using this pattern the teacher may explore students' ideas through dialogic interaction (Mortimer & Scott, 2003).

Both teacher initiation and feedback moves are important in classroom dialogue. Teacher initiation is typically studied by analysing the form of questions. If the teacher asks open-ended or authentic questions, she has the opportunity to get information from student thinking, whereas closed questions typically lead to the before-described IRE pattern (Chin, 2007; Dysthe, 1996). Teacher feedback can be in the form of affirmation (neutral rather than evaluative) or instruction, but it also can be a follow-up question (Chin, 2007; Oliveira, 2010; Myhill & Dunkin, 2005; van Zee & Minstrell, 1997; Dysthe, 1996). The purpose of these open questions is to help students elaborate their thinking and construct their conceptual knowledge. There is strong evidence for the benefits of teacher questioning skills and proper feedback for student learning (Hattie & Timperley, 2007).

The studies of on-the-fly assessment have been informed by the research of classroom dialogue described above. When teachers implement formative assessment practices in classrooms, they (amongst other things) elicit, interpret and use data from students to make decisions about the next steps for learning (Black & Wiliam, 2009; Bell & Cowie, 2001). In on-the-fly formative assessment, eliciting, interpreting and using data are part of the teacher–student conversation. For exploring these elements in teachers' on-the-fly practices, Ruiz-Primo and Furtak (2006) developed the ESRU framework. Instead of the IRE or IRF pattern, the ESRU model contains four steps: (1) the teacher elicits (E) information from the student by formulating a question, (2) the student responds (S), (3) the teacher recognises (R) the student's response and (4) finally, the teacher uses (U) the collected information to support student learning. The complete ESRU cycle contains all these steps. Ruiz-Primo and Furtak (2006; 2007) reported positive relations between student achievement and teachers' tendency to use complete ESRU cycles. Recently, the ESRU model has been applied and advanced by its developers (Furtak et al., 2016; Furtak et al., 2017), for example, by combining the codes' 'recognising' and 'using' from the original ESRU model.

Forms and functions of on-the-fly assessment conversations

Differentiating between the forms and functions of actions or objects is common in arts and science. In the context of classroom assessment, the form of assessment may refer to the means used to get information from students' performance and to evaluate it, such as the use of an open-ended problem, a rubric or a portfolio (Saxe, Gearhart,

Franke, Howard, & Crockett, 1999; Hobson, 1997). These forms of assessment always serve some functions. For example, in a mathematics class, a form of assessment can be using an open-ended problem to elicit students' performance, where the function can be to provide the teacher an opportunity to observe students' problem-solving methods and their conceptual understanding (Saxe et al., 1999).

In classroom talk, which includes on-the-fly formative assessment conversations, form may refer to the phonology and syntax of talk, and function to its communicative and/or cognitive purposes (Westgate & Hughes, 1997). Myhill and Dunkin (2005) applied a grounded theory approach to study teacher questions in whole-class discussions. They reported four forms ('factual', 'speculative', 'process' and 'procedural') and eleven different functions of questions. 'Speculative' and 'process' forms of questions, which often invite higher-order thinking, were infrequent, while 'factual' questions ('questions that invited a predetermined answer') were most common (64%) in their data. Some of the factual questions (26%) related to the function 'factual elicitation', but many of them related to other functions, such as 'building on content', 'building on thinking', 'checking prior knowledge' and 'developing reflection'. Thus, Myhill and Dunkin (2005) showed that the form of the question cannot determine its function.

In this study, we utilise the conceptualisation of forms and functions of classroom assessment (Saxe et al., 1999) and teacher questions (Myhill & Dunkin, 2005; Oliveira, 2010; Westgate & Hughes, 1997) in the context of on-the-fly assessment conversations. We consider forms of on-the-fly assessment to mean the ways teachers collect information related to the students' learning (e.g. using closed questions), and how they use that information (e.g. giving instructions). Further, in this study, functions of on-the-fly assessment refer to the purposes of the formative assessment in episodes (e.g. helping students to articulate their thinking).

The research gap and the research questions

Several previous studies have focused on the on-the-fly assessment and helped the research community to understand this complex phenomenon. This study attempts to increase this understanding by using different methodological choices. First, previous studies have mostly analysed teacher-student interaction on the level of speaking turns, while this study takes a wider perspective and uses the whole conversation (episode) as a unit of analysis. One consequence of this decision is that interpreting a meaning of an action (e.g. a teacher move) can be based on many speaking turns, and unspoken acts as well. For example, when analysing speaking turns, teacher eliciting is usually considered as *teacher questioning* (Ruiz-Primo & Furtak, 2006; Oliveira, 2010), although it is known that students' questions and initiations play an important role in classroom discourse (e.g. Louca, Zacharia, & Tzialli, 2012; Chin & Brown, 2002). In our previous study (Correia et al., 2016), we found that students actively asked for help and made other initiative statements that could be a source of information for the teacher without active eliciting by questioning. To illustrate this, we refer to an excerpt of a discussion presented by Westgate and Hughes (1997, p. 134). In the excerpt, a student says 'Ice is hard not soft like snow "cos if you slip it hurts"', which is coded as student initiation. Then the teacher says 'Have you ever felt ice?', which is coded as teacher initiation and eliciting. Another interpretation could be that the teacher's utterance is not eliciting but actually a follow-up question in which the teacher is using information gleaned from *listening* to what the student said. Thus, considering means of eliciting other than questioning can change the interpretation. For this reason, it is important to analyse how teachers gather information related to students' thinking using different means. Since the term elicit often relates to teacher questioning, we will use the term *collect* in this paper.

The second big difference from previous on-the-fly studies is analysing the form and function of on-the-fly assessment conversations. Types of collecting and using, as well as the purposes of formative assessment, have been studied, but these have not been conceptualised as forms and functions, and have not been analysed in parallel for the same unit of analysis. Differentiating between forms and functions brings a novel perspective to our on-the-fly study and opens up new possibilities. Considering forms and functions separately may result in recognising several types of forms and functions. Furthermore, functions could relate to the flow of student inquiry, since teachers may have different aims depending on the students' current situation. Additionally, the forms and functions may help in recognising differences between lessons.

Third, on-the-fly assessments have usually been studied in the context of whole-class discussions; however, our own studies have focused more on small-group discussions (Nieminen, Hähkiöniemi, Leskinen, & Viiri, 2016; Correia et al., 2016; Harrison et al., 2018). Ruiz-Primo and her group have recently studied small-group work, but the results are not yet published (oral presentations referred to in Furtak et al., 2017). In this paper, we focus mainly on the teacher interacting with a small group of students (2–3) who are conducting a physics inquiry, albeit the data include a couple of episodes from whole-class discussions.

The study is framed by the three research questions below. The first and second research questions guide the process by which data-driven categories were created. The third research question serves our attempt to show the potential of forms and functions in distinguishing teachers' formative assessment practices.

(1) How can the forms of on-the-fly assessment be described by interpreting how a teacher collects and uses the information?
(2) How can the functions of on-the-fly assessment be described in relation to the flow of student inquiry?
(3) What differences between two lessons can be identified by analysing forms and functions of on-the-fly assessment?

Methods

Data collection

Data were collected as part of a project in which teachers implemented physics and mathematics inquiry activities with the aim of having on-the-fly formative assessment discussions with the students. Voluntary teachers (3 physics and 5 mathematics) were found by personal contacts and asking rectors from one middle-sized Finnish city. In this study, we focus on two teachers' practices who implemented the same activity in 7th grade (13–14 years old) physics classes. The activity and the lessons are described below. Both teachers had several years of teaching experience. In this paper, we use pseudonyms Smith and

Brown for the teachers. Regarding the topic of the study, we do not see it as necessary to indicate the gender of the teachers and will use the pronoun 'she' regardless of the gender.

The researchers planned the activity and included some recommendations for possible use of on-the-fly assessment in the lesson. The lessons were video-recorded with one camera following the teacher from the back of the class. The camera was connected to a wireless microphone on the teacher. The microphone also captured the students' utterances when they spoke to the teacher. In addition, students' written productions were collected. The participants and the students' parents were informed about the study. They voluntarily accepted to take part and returned signed research permits.

The inquiry activity and the structure of the lessons

The type of inquiry activity was predict-observe-explain (White & Gunstone, 1992), which is suitable for formative assessment since students have opportunities to explain their thinking (Furtak et al., 2016). The activity included three tasks for the image formation on a plane mirror. Inquiry was structured so that the students followed ready posed questions and methods (Bell, Smetana, & Binns, 2005). An example of a task is presented in the Appendix. In this task, the mirror is covered by a cloth. Students need to predict who can see (A or B) the image of an object if the cloth is removed. After their prediction, they study the issue with a real plane mirror. Then they are asked to explain how their prediction and observation differ. Tasks were adapted from Goldberg and McDermott (1986).

Table 1 presents the phases and times used for the inquiry activity in the classes. First, Smith showed a video about the analogy between a bouncing billiard ball and reflection of light. Then she presented the law of reflection as a part of the introduction for the activity. Brown just stated that the homework was to read a chapter concerning the plane mirror and showed the same video as Smith. Smith used more time for the introduction as she presented the theory of light reflection.

In both lessons, students worked in small groups or pairs. Brown's class spent about twice as much time for the group work phase. Brown did not need time for classroom management as the students focused on the task well, but Smith needed to put pressure on a couple of students.

Smith used 10 min for closing, which was mainly teacher monologue. There was only a short discussion, which included on-the-fly assessment. Brown had a long discussion (13 min) with students based on their findings and alternative conceptions.

Generally speaking, Smith used less time for the inquiry activity. She started from the theory whereas Brown presented the theory afterwards. Smith's lesson was more hectic as

Table 1. Events and times regarding inquiry activity.

Teacher	Introduction	Group work	Closing	Total time
Smith	10 min: The billiard video and the teacher's introduction for the theory of reflection	17 min: The teacher circulates in the groups (on-the-fly)	10 min: Teacher presents correct answers. One short discussion (on-the-fly)	37 min
Brown	5 min: the billiard video and a discussion about it (on-the-fly)	32 min: The teacher circulates in the groups (on-the-fly)	13 min: Teacher discusses with students about findings (on-the-fly)	50 min

some students were restless. Smith proceeded in a rapid pace and used more time for classroom management, albeit most of the students worked well.

In relation to the inquiry model of Pedaste and others (2015), our introduction phase relates to the 'Orientation' and the group work phase overlaps the phases of 'Conceptualization', 'Investigation' and 'Conclusions'. Further, in this paper, on-the-fly interactions are analysed concerning various sub-phases of inquiry, which differ in terms of content and duration. A sub-phase can be, for instance, making a prediction in Task 1, observing a certain issue in Task 2 or trying to write down observations in Task 3. A sub-phase also can be a part of the whole class discussion when the teacher is discussing with students concerning a certain issue such as an alternative conception that had arisen during group work.

Data analysis

At first, the lesson videos were divided into on-the-fly episodes in which the teacher discussed with students issues related to a sub-phase of the inquiry. An episode ends when the teacher–student interaction stops (e.g. the teacher moves to another student group), or the sub-phase of students' inquiry process changes to a new sub-phase, for example when the theme of discussion changes during the closing discussion. In total, 37 on-the-fly episodes were identified and transcribed. Both videos and transcripts were analysed side by side since gestures, gaze direction and use of objects such as a mirror or pen were important in addition to talk.

After selection of episodes, they were analysed through data-driven analysis (Corbin & Strauss, 2008; Miles & Huberman, 1994). At first, we examined the episodes and characterised how the teacher collects information of the students' learning and how she uses that information. In doing this, we realised that our interpretation of collecting depends on our interpretation of using. Thus, we decided to interpret collect-use dyads from an episode as a whole instead of coding speaking turns. When we had interpreted preliminarily that the teacher had gotten information of student learning, we examined whether the teacher used that information. If the teacher did not use the information, then we changed our preliminary interpretation so that the teacher was not actually collecting. Similarly, when we had a preliminary interpretation that the teacher used information to guide the students, we searched how she got the information upon which the guidance was based. Through recognising different collect and use types as well as comparing and refining the types, we composed a set of five collect and four use types. These collect and use types described the form of the on-the-fly assessment discussions. If an episode contained two collect or use types, we interpreted which one of these was more dominant. Thus, each episode always includes just one collect-use dyad.

After analysing the forms of the on-the-fly assessments, we interpreted the function of the on-the-fly assessment in supporting student learning in each episode. After making a tentative interpretation, we compared it to the tentative interpretations of the other episodes and, if needed, refined the interpretations. Through this, we created a set of categories for the episodes based on their interpreted functions. The categories were examined for similarities and grouped into main categories (such as F1 in Table 3). Some categories were also merged. Finally, we had created four main categories (F1–F4) with nine subcategories (e.g. F1a).

Although the analysis is qualitative, we counted some frequencies. The purpose of examining the frequencies was to notice and understand qualitative differences between the two lessons. The analysis was highly interpretative, and we did not try to generalise our results or create a general coding manual. Thus, we did not calculate inter-rater agreement. Instead, the analysis was done by two researchers (investigator triangulation). All the episodes were discussed carefully to achieve consensus about meanings of utterances and gestures, and eventually, categories of the forms and functions.

Results

In the first section, the results for the first and second research questions are presented. The text is organised according to the main types of functions. Then, the results for the third research question follow in the second section.

Forms and functions of on-the-fly assessment

To help the reader grasp the overall idea for the results, all the forms and functions of on-the-fly assessment are presented in Tables 2 and 3. Table 2 shows five different types of how the teachers collected information from students' thinking (C1–C5) and four types of how the teachers used that information (U1–U4). It must be noted that collect and use questions are defined similarly because the question can be used to collect information or it can already be based on collected information. The forms are described below with the numbered extracts.

Table 3 shows four main types of functions and their sub-types. The functions and their differences are described in the next sections.

Function 1: Checking or examining thinking

In the episodes belonging to this main category, the teachers got information about the students' learning but did not guide the students to develop their ideas further or to

Table 2. Types of collecting and using interpreted from the on-the-fly episodes.

Types of collecting and using	Description	Extracts
C1 Collect by open question	Inviting a response that is not known in advance and using the achieved data later.	1, 2, 3, 8, 10
C2 Collect by closed question	Inviting a predetermined answer and using the achieved data later.	9
C3 Collect by listening	Listening to what students are saying to each other or to the teacher even though the teacher is not questioning. The achieved data are used later.	6
C4 Collect by observing	Observing what students are doing or what they have written without questioning or students talking to the teacher. The achieved data are used later.	4, 5, 7
C5 Collect by previous finding	Achieving data from students before the episode and using the achieved data straight away in the beginning of the episode.	11
U1 Guide by open question	Inviting a response that is not known in advance based on already collected information.	3
U2 Guide by closed question	Inviting a predetermined answer based on already collected information.	4, 5, 7, 9
U3 Direct instructions	Giving instructions what to do based on already collected information.	2, 6, 8
U4 Confirm	Letting students know that they are doing good job. Not guiding. Action is based on already collected information.	1, 10

Table 3. Functions of on-the-fly assessment interpreted from the on-the-fly episodes.

	Functions of on-the-fly	Extract
F1	*Checking or examining thinking*	
F1a	Checking quickly that students are on the right track	1
F1b	Checking students' written work is adequate	2
F1c	Finding how students are thinking and helping to articulate their ideas	3
F2	*Correcting or developing thinking*	
F2a	Correct student thinking based on teacher thinking	4
F2b	Correct student thinking based on student thinking	5
F2c	Develop student ideas based on student thinking	6
F3	*Redirecting thinking*	
F3a	Focus student attention on an issue and activate thinking	7
F3b	Help to begin explaining	8
F3c	Help to understand task or start solving the task	9
F4	*Review thinking (set up or sum up)*	10, 11

build new ideas. Rather, the teacher just acknowledged how the students were proceeding and possibly helped them in articulating the ideas that they already had developed. Some of the episodes were short exchanges but some were longer conversations.

Just before the first extract, students were talking to each other about how they observed the location of an image on a plane mirror. The teacher Smith stood next to the students and listened.

Extract 1

Turn	*Speaker*	*Transcription*
1	Smith	Where's the image?
2	Student 1	It can be seen on a different location, but it does not move.
3	Student 2	But it is in the middle.
4	Smith	Yeah, just like that. Mark (write down) that location.

After having listened to the students' discussion, Smith asked about the image. Students told their observation and Smith confirmed that it is correct. She also asked them to write the observation down, but this is not considered as guiding because it was just repeating the task. In this episode, Smith got information of students' learning by asking an open question and used this information to confirm the students were right. Thus, the form of the on-the-fly assessment in the episode is characterised by the following collect and use types: collect by an open question (C1) and confirm (U4). In this episode, the open question and confirming are both in their own speaking turns (1 and 4), but generally, the types of collecting and using were not decided based on single speaking turns. For example, excerpt 4 shows well how the interpretation of the types is based on the whole episode.

In Extract 1, we interpreted that the teacher intended to have a snapshot of how the students were proceeding and saw no reason to interfere more with their work. In other words, the function of the on-the-fly assessment in this episode is to *check quickly that students are on the right track* (F1a).

The second extract started when teacher Brown came to a student group. The students had written down their findings about the area where an observer can see an object via a plane mirror.

Extract 2

Turn	Speaker	Transcription
1	Student	We got it now … this result.
2	Brown	You got the result. Yeah. What was this line?
3	Student	That was the area of non-seeing.
4	Brown	Also mark that. Mark what these areas mean.

The students said that they had a result, but just from this utterance Brown did not get notable information about their results. Thus, Brown asked an open question about the meaning of their notation. The student answered, and Brown instructed the students to complete their written work by writing down the meaning of their notation. Brown got information through an open question and gave instructions based on the achieved information. Thus, the form of the on-the-fly assessment in the episode is to collect by an open question (C1) and use by direct instructions (U3). In this extract, the function of the on-the-fly assessment is similar to the function in Extract 1. The difference is that in Extract 2 the teacher checked that the students' written work is adequate when in Extract 1 the discussion focuses on doing observation. Thus, we interpreted that the function of the on-the-fly assessment in extract 2 is *checking students' written work is adequate* (F1b).

In Extract 3, Brown came to a pair of students. She opened the discussion.

Extract 3

Turn	Speaker	Transcription
1	Brown	What is going on here?
2	Student 1	Well, when … if we put that … that when …
3	Brown	Are you working with conclusions?
4	Student 1 & 2	Yeah.
5	Brown	Okay. Do your prediction and your observation differ?
6	Student 1 & 2	Yeah.
7	Brown	In what way?
8	Student 2	We thought as if I sit here. Then the mirror is there, so it would be mirrored to this corner, but it is in that corner.
9	Brown	Hm. So, you were here. Here is the object. You sat here. And you see the image … you supposed that it is here, but it is on this side.
10	Student 2	Yes.
11	Brown	Well. So, you can answer: Do your prediction and observation differ?
12	Student 2	Yeah.
13	Brown	Yes. How does the person's location influence the location of the image?
14	Student 2	It is mirrored like … If you look from the side the object is mirrored to that corner where you sit or stay.
15	Brown	To that corner? So, you considered that the image is like on the different side from the object than your eye, but it is on the same side …
16	Student 2	Yeah.
17	Brown	… as your eye. Okay. Yes it is. Yes. Then just write.

In Extract 3, Brown asked many questions to collect information. She used the information to instruct students by revoicing student responses (turns 9 and 15) and to ask an open question (turn 13). Thus, the form of the on-the-fly assessment in the episode is to collect by an open question (C1) and guide by an open question (U1). Although Brown gave input in the discussion, her inputs encouraged students to explain their ideas more thoroughly rather than develop their ideas. Therefore, in this episode, the function of on-the-fly assessment is *finding how students are thinking and helping to articulate their ideas* (F1c). In this episode, the teacher does not try to correct student thinking.

Function 2: Correcting or developing thinking

The common characteristic in the episodes belonging to this main category is that the teacher got information of the students' thinking and then, based on the achieved information, tried to correct or develop the students' thinking.

In Extract 4, students were considering the place of an image on the plane mirror. The teacher Smith approached the students.

Extract 4

Turn	*Speaker*	*Transcription*
1	Smith	Does the image appear exactly on the mirror surface (*points to the students' paper, then the mirror surface*) or a little like behind the mirror (*pointing behind the mirror*)?
2	Student 1	Well, when that no …
3	Smith	Where is the image?
4	Student 1	That is as moving there … (*The Student's explanation continues. He explains that the image moves when looking from different sides.*)
5	Smith	But is the image now on the surface of the mirror or behind the mirror?
6	Student 2	On the surface of the mirror.
7	Smith	Is it so?
8	Student 2	I don't know.
9	Smith	Is your image now here on the mirror surface?
10	Student 2	No, it is behind.
11	Smith	Behind. Yes. (*Goes away.*)

When Smith came to the students, she started to speak almost immediately. Smith probably observed that the students might think that the image is located on the surface of the mirror. Already her closed question in turn 1 is based on the information gained through the observation. In turn 3, Smith asked an open question, and Student 1 explained his ideas about the location of the image in the plane parallel to the mirror. However, Smith ignored these ideas and returned the discussion to the track she wanted, that is to thinking about the location of the image in the plane that is perpendicular to the mirror (closed question in turn 5). Because the guidance was not based on the information gotten through the open question, the open question does not determine the type of collecting. Instead, the form of the on-the-fly assessment is to collect by observing (C4) and guide by closed question (U2). As regards the function, the teacher quickly got some information of the students' probable thinking but interpreted this strictly from the teachers' own perspective and tried to correct the students' thinking. Thus, the function of the on-the-fly assessment is to *correct student thinking based on teacher thinking* (F2a).

In Extract 5, Brown came to a student group. She looked at the students' written work and started a discussion with them.

Extract 5

Turn	*Speaker*	*Transcription*
1	Brown	Is that the area, right?
2	Student 1	Yes.
3	Brown	I have to ask, did you have your eye between this object and the mirror?
4	Student 2	(It is) not between them.
6	Brown	Be careful.
7	Student 2	So … Do I have to draw from a long distance?
8	Brown	How do you want to … but I know that you did not have your eye here between, didn't you? Yeah. That is, that the eye has probably been somewhere in this area.
9	Student 1	… there … (inaudible)
10	Brown	Yes.

Most likely, Brown observed an erroneous placement of the observer and the object in the students' paper. Her closed question in turn 3 helped the students to notice their error. Thus, the form of the on-the-fly assessment is collect by observing (C4) and guide by closed question (U2). Extract 5 has commonalities with extract 4. However, in extract 5, the teacher guided the students only to notice their mistake without guiding them further based on the teacher's own view. Thus, the function of the on-the-fly assessment in extract 5 is *correct student thinking based on student thinking* (F2b). Extracts 4 and 5 illustrate that although the form of the on-the-fly assessment may be the same, the functions may be slightly different.

Extract 6 started when a student called Smith to their group. The student started a discussion.

Extract 6

Turn	*Speaker*	*Transcription*
1	Student 1	If I have a mirror and a pen here ...
2	Student 2	It does not move at all.
3	Student 1	Yes, that is not moving. When I look from here, it is in the same place.
4	Smith	Correct.
5	Student 1	So, it is somewhere here.
6	Smith	Yes, it is there behind, the pen. As far as it is ... This is here in front and the image is somewhere behind.
7	Student 1	So, can I just put it here?
8	Smith	Yes. Certainly, as far as there. So, isn't it as far as this object? So, the image is as far?
9	Student 1	Yes.

The students explained their findings and Smith collected by listening (C3). Smith echoed the students' ideas but also completed their finding (turns 6 and 8), which can be seen as giving instructions (U3). In this episode, the teacher was not trying to correct the students' ideas but caught at their idea of the place of the image remaining the same even though it is observed from different places. Smith tried to develop their idea further by making it explicit that the image is as far from the mirror as the object. Thus, we considered the function of the on-the-fly assessment being to *develop student ideas based on student thinking* (F2c).

Function 3: Redirecting thinking

In the episodes of this main category, the teacher first achieved information of where the students were or how they were thinking. Then, the teacher guided them to transfer to a different sub-phase of the activity or to think about a different issue than what they were thinking.

In Extract 7, Smith came to a pair of students and looked at their paper. A student expressed they had made their hypotheses for the plane mirror task.

Extract 7

Turn	*Speaker*	*Transcription*
1	Student	We are ready.
2	Smith	If you look at the mirror image, does it show that you are on the surface of the mirror or behind the mirror?
3	Student	Behind the mirror.
4	Smith	Well, look again (*giving students the mirror*). You have marked the image place here on the surface of the mirror (*pointing to the students' paper*).

Evidently, Smith collected by observing (C4) the students' papers because, in turn 4, she expressed having noticed that they had marked the image into the mirror. Based on the observation, Smith guided the students mainly by a closed question (turn 2) together with some instructions (turn 4). In this episode, the discussion highlighted that the students said a different hypothesis than what they had written on the paper and that they had to examine this further with the mirror that Smith gave to them. The teacher did not correct the students' thinking or tell which one of their responses (written or orally stated) is correct but transferred them to investigate the issue with a mirror. Therefore, the function of the on-the-fly assessment is to *focus student attention on an issue and activate thinking* (F3a).

Extract 8 started when Smith came to some students. She initiated a discussion.

Extract 8

Turn	*Speaker*	*Transcription*
1	Smith	What did you investigate?
2	Student 1	Yes. A sees, but B does not.
3	Smith	Fine. How do you now explain this?
4	Student 1	Well, it is so hard to explain. That (*inaudible*) sees, but that does not see.
5	Smith	If you would draw light rays there. How is the ray reflected from that object to the mirror?
6	Student 2	(*Explains something that is inaudible.*)
7	Smith	Hmm … Draw something. (*Goes away.*)

Extract 8 began when Smith collected information with an open question (C1). Through this, the teacher got information on what kind of observation the students had made and that they had difficulties taking the next step, that is to explain their observation. The teacher instructed them to use the ray model (U3). The teacher got information where the students were and helped them to transfer to explaining their observation by drawing. Thus, the function of the on-the-fly assessment is to *help to begin explaining* (F3b).

In Extract 9, Smith came to a group and started to place the equipment for the students.

Extract 9

Turn	*Speaker*	*Transcription*
1	Smith	At least you can investigate how it happens. Fix the object now then … mirror … object is here a little aside. (*Teacher arranges the equipment.*) Ok, when you look from there what do you see?
2	Student 1	Well, I don't know how much it has to be here or how far it has to be, so that it depends on that.
3	Smith	Well, does Student 2 see it?
4	Student 2	What?
5	Smith	This eraser?
6	Student 2	No.
7	Smith	Do you see (Student 1)?
8	Student 1	Yes I do.
9	Smith	Well, now you can write down the location where you see it. (*Gives advice to do the inquiry.*) Or, what does Student 2 think? Behind the surface of the mirror or on the mirror?
10	Student 1	Behind (quietly).
11	Smith	How does the image look? Where is it?
12	Student 2	On the mirror.
13	Smith	The image is on the mirror, or behind the mirror?
14	Student 2	Ah, well, if you think that way then it is behind it.
15	Smith	Yeah.
16	Student 2	I did not think that it is there, because it is not.
17	Smith	The place of the image. Where the image seems to be. It was the question.
18	Student 2	Well, behind the mirror.
19	Smith	Yes.

It may be that Smith had already observed that the students had difficulties in understanding the task because Smith started to place the mirror and the object for them. In any case, she also asked a closed question (turn 1), which revealed that the students had problems understanding the task. Thus, Smith asked a closed question (C2) to collect information. After this, she guided them with many closed questions (U2). Smith helped the students to understand the task and continue from there. Thus, the function of the on-the-fly assessment is to *help to understand the task or to start solving the task* (F3c).

Function 4: review thinking (set up or sum up)

The descriptive feature of this main category is that the teacher is reviewing the students' thinking or comprehension. The purpose is to prepare students for the forthcoming activity or to have them reflect on the already completed activity.

Extract 10 is from the introductory phase of the lesson. Brown had just shown the students a short video featuring an analogy between the bounce of a billiard ball from the wall and the reflection of a light ray. She initiated a discussion on the video clip.

Extract 10

Turn	*Speaker*	*Transcription*
1	Brown	How is this related to the mirror? How is billiard playing related to mirrors?
2	Student 1	The ball bounces from there ... (*inaudible*)
3	Brown	From the cushion? Yes.
4	Student 1	... back. (*Inaudible.*)
5	Brown	Yes. What do you think, Student 2?
6	Student 2	It means that if the ball would be light, so, the side would be the mirror. It reflects it always as that ... What is that?
7	Brown	Student 3?
8	Student 3	It goes in the same angle to the other direction. When it comes, it goes out in the same angle.
9	Brown	Yes, ok. And this was our short introduction to the topic.

Brown asked the students' comprehension of the analogy (collect by open question, C1) and, in the end, just accepted their explanation (confirm, U4). Her intention was to set up the forthcoming inquiry activity about the image formation on the plane mirror. Thus, the function is *review thinking* (F4) and in particular to set up the activity.

The shortened Extract 11 is part of the closing discussion after the students' inquiry. A student had expressed an alternative conception during the group work. Brown launched a discussion on this.

Extract 11

Turn	*Speaker*	*Transcription*
1	Brown	Another issue is how does the mirror form the image ... (*shortened*). Where does the light start? I have to ask you, Student 1?
2	Student 1	I don't know.
3	Brown	What do you think? Guess?
4	Student 1	From the eye.
5	Brown	From the eye. Hmm ... Do you remember what was the object's ... If you see an object, for instance this dishcloth, two possibilities when you can see an object? What were those? How is it related to light? Student 2?
6	Student 2	Well, you can see it when light is reflecting from it.
7	Brown	This dishcloth is seen when light is reflecting from it or ... Student 2?
8	Student 2	It emits light.
9	Brown	It emits light. Which one is it? Does this dishcloth emit light or is light reflected from the dishcloth? Student 3?

(*Continued*)

Continued.

Turn	*Speaker*	*Transcription*
10	Student 3	Reflected.
[The discussion has been shortened 11 speaking turns in order to shorten the text.]		
22	Brown	I still want to ask you again, Student 1: Where does the light ray start?
23	Student 1	From the object.
24	Brown	From the object, not from the eye. This is a rather common idea that light starts from the eye, but we have two receivers. Hey! Does sound start from the ear? (*Students shake their heads.*) No. The ear is a receiver; with the ear you observe that. At this moment, sound starts from my pharynx, and you hear it because you have the receiver. In the same way light comes to this classroom, and you see this dishcloth because light is reflected from the dishcloth to your eyes; the eye is the receiver.

Brown had observed a misconception during the group work. Before asking the question in turn 1, Brown already knew that the misconception existed. Thus, this question does not determine the type of collecting. On the contrary, Brown collected information by the previous finding (C5). After this, the questions she used can be considered mainly as closed questions because her intention was to receive certain answers (guiding by closed question, U2). In Extract 11, the teacher looked backward and wanted to raise issues from the group work phase to support learning. Thus, the function of the on-the-fly assessment was to *review comprehension* (F4) and particularly to sum up the activity.

Differences between the lessons in forms and functions

The distributions of types of collecting in the two lessons are presented in Table 4 and distributions of types of using in Table 5. It seems that there were some small differences, but nothing points to a dramatic difference between the lessons. Collect by observing was more typical of Smith in this lesson. She often went to student groups looking at their papers and started to guide them based on a quick interpretation of the students' needs. This may also relate to Smith's lesson having less time for group work and to the hectic atmosphere in the lesson. Collecting by previous finding occurred only in Brown's final discussion when she summed up students' findings. In summary, based on the forms of the on-the-fly assessment, no substantial differences between the lessons could be observed.

The distributions of the functions of the on-the-fly assessment in the two lessons are presented in Table 6. Both teachers had instances in all the main categories, but there were differences in the subcategories. For example, F1c was quite typical for Brown, and these discussions tended to be quite long, as in Extract 3. Instead, type F2a was present only in Smith's lesson when she guided students to observe that the image was located behind the mirror and not on the surface. Furthermore, in Brown's lesson many of the functions belonged to the categories F1c, F2b and F2c, which are based on student thinking. This may be related to Brown devoting more time to group work and proceeding at a slower pace. F2c and F3c were common types for both the teachers. Brown's lesson contained more of the function review comprehension (F4) before (set up) and after (sum up) student inquiry. In summary, examining the functions of the

Table 4. Frequency and percentage (parenthesis) of the types of collecting.

Teacher	Collect by open question	Collect by closed question	Collect by listening	Collect by observing	Collect by previous finding	Total
Smith	7 (43.8)	1 (6.3)	5 (31.3)	3 (18.6)	–	16 (100)
Brown	12 (57.1)	–	6 (28.6)	1 (4.8)	2 (9.5)	21 (100)

Table 5. Frequency and percentage (parenthesis) of the types of using.

Teacher	Guide by open question	Guide by closed question	Direct instructions	Confirm	Total
Smith	3 (18.8)	6 (37.5)	5 (31.3)	2 (12.5)	16 (100)
Brown	5 (23.8)	4 (19.0)	7 (33.3)	5 (23.8)	21 (100)

Table 6. Frequencies and percentages (in parenthesis) of the functions of the on-the-fly assessment in the two lessons.

	Function	Smith	Brown
F1	*Checking or examining thinking*		
F1a	Checking quickly that students are on the right track	2 (12.5)	–
F1b	Checking students' written work is adequate	–	5 (23.8)
F1c	Finding how students are thinking and helping to articulate their ideas	–	4 (19.0)
F2	*Correcting or developing thinking*		
F2a	Correct student thinking (based on teacher thinking)	3 (18.8)	–
F2b	Correct student thinking (based on student thinking)	–	2 (9.5)
F2c	Develop student ideas (based on student thinking)	3 (18.8)	3 (14.3)
F3	*Redirecting thinking*		
F3a	Focus student attention to an issue and activate thinking	1 (6.3)	–
F3b	Help to begin explaining	2 (12.5)	–
F3c	Help to understand task or to start solving the task	4 (25.0)	3 (14.3)
F4	*Review comprehension*	1 (6.3)	4 (19.0)
Total		16 (100)	21 (100)

on-the-fly assessment seemed to reveal some differences in the nature of formative assessment in these two teachers' lessons.

Discussion

An essential part of teachers' formative assessment is eliciting information from students and using the information to support learning (Black & Wiliam, 2009; Bell & Cowie, 2001). We were interested to study the forms and functions of these important aspects of formative assessment in real-time, teacher–student assessment conversation called on-the-fly formative assessment.

Forms of on-the-fly assessment

As discussed in the introduction, many previous studies have analysed teachers' on-the-fly interactions from single speaking turns (Ruiz-Primo & Furtak, 2006, 2007; Sezen-Barrie & Kelly, 2017). In addition, we noted that teacher eliciting is usually understood as teacher questioning. We confronted also some problems with speaking turn level coding in our previous studies (Nieminen et al., April, 2016; Harrison et al., 2018), especially to differentiate E and U components in the ESRU framework. Another problem consisted of episodes without a speaking turn including teacher eliciting. In these episodes, which are typical in Finnish classes, a student initiated the episode. Then, a no speaking turn was coded as eliciting, and it looked like there is no teacher eliciting. In this study, we solved these problems by examining an on-the-fly interaction episode as a unit. This more holistic view allowed us to see the form by which the teacher actually collects information and how she uses that information. In addition, episode-level coding allowed us to better capture the reciprocal connection between collect and use as the type of each teacher's collect and use were determined together even though they received separate codes.

As collecting was not coded for speaking turns, we also found other types than question-based eliciting. In total, we found five different types: collect by (1) open questions, (2) closed questions, (3) listening, (4) observing and (5) previous findings. Collecting by questions was the most numerous type (20 episodes), but other types (17) were common as well. Thus, the teachers used various modes to collect information from students. Collecting by open and closed questions, that is teacher eliciting with questions, has been studied a lot (e.g. Oliveira, 2010; Chin, 2007; Ruiz-Primo & Furtak, 2006), but other modes to collect information have not been reported widely. One rare example is a conference paper by Ruiz-Primo and Kroog (2017). They presented a multidimensional observation protocol including types of collecting such as asking questions, listening to students and walking around to observe students' work.

Another important aspect is that a certain action in the beginning of an episode does not necessarily determine the type of collecting. Often, the teacher may initiate the episode by asking a question, but it does not define the type of collecting. For example, in Extract 5 the teacher asked a question, which may resemble collecting. However, this question was a part of 'using' because she had already collected information from the students' written work by observing, and the first question was based on this. In some cases, students may initiate the episode, but it does not mean the teacher collects enough information by listening. For example, in Extract 2, students said that they had gotten a result, but the teacher did not use only this information. Instead, the teacher both observed students' written work and asked about their results. Then, based on this, she asked the students to complete their answer.

As in the case of collecting, the type of use was determined based on the dominant feature of the use of information. For example in Extract 4, the teacher asked an open question and got information from the students' thinking, but she did not use that information. Instead, she continued asking closed questions, which determined the type of using. Generally speaking, the teachers used information by asking closed and open-ended questions, for example, to direct the students' thinking or to probe their understanding. They also gave direct instructions and confirmed the students' thinking. These types of 'using' are similar to those previously reported (e.g. Harrison et al., 2018). Indeed, open, authentic questions are important for dialogic interaction where students' ideas can be heard and formatively assessed (Chin, 2007; Dysthe, 1996; Mortimer & Scott, 2003), but also closed questions and more authoritative talk are needed to guide learning in a certain direction (Scott, Mortimer, & Aguiar, 2006; Lehesvuori, Viiri, Rasku-Puttonen, Moate, & Helaakoski, 2013).

In our approach, the episodes were coded as collect-use dyads. Although this decision had above-explained affordances, it also had a constraint. Our coding system does not take into consideration episodes in which use is missing as in incomplete ESRU cycles (Ruiz-Primo & Furtak, 2007). Thus, our coding system does not allow the study of possible missed opportunities for formative assessment (Constantinou & Papadouris, 2016).

Functions of on-the-fly assessment

Analysing only the form of the on-the-fly assessment did not capture the essence of the assessment practice, but we needed to also consider the function of the on-the-fly assessment. This is similar to the study by Myhill and Dunkin (2005), who noticed the need to study both form and function of teacher questions, which is also similar to the reformulation of telling based on function instead of form in Lobato, Clarke, and Ellis (2005).

To understand the functions of on-the-fly assessment without any predefined theoretical categories, data-driven analysis was conducted for all the episodes from the two teachers' lessons. The following main categories for the functions were found: (F1) Checking or examining thinking, (F2) Correcting or developing thinking, (F3) Redirecting thinking and (F4) Reviewing thinking (set up or sum up). The basic idea, which separates the main types (F1–F4) from each other, is the teacher's objective in relation to the students' sub-phases of their inquiry process. Figure 1 illustrates the hypothetical sub-phases (the rectangles) and functions (the arrows and magnifiers). Here, the sub-phases do not follow any pre-planned learning progressions (some of them could be), but they are any kind of phases that a student group encounters in their inquiry process. These sub-phases and their number may differ between student groups. For example in Extract 8, the student group has made an observation (their present sub-phase). The next sub-phase would be to explain it in written format, and the teacher helps them to get started (a tip about the ray model), that is, helps them to take the step on to the next sub-phase.

In Function F1 (the magnifiers in Fig. 1), the teacher's objective is to check or examine the students' thinking or working, but not to guide the students to develop their ideas further or to build new ideas. Thus, F1 is illustrated as a magnifier focused on the students' work. Typically, the teacher just checks what the students are doing and maybe helps them to express the ideas that they already have.

Instead, Function F2 includes on-the-fly episodes in which the teacher corrects or helps students develop their thinking. Here, the students do not proceed to the next sub-phase, but they are improving their present idea with the help of the teacher. In other words, the teacher aims to change their thinking but inside the sub-phase they are in. Thus, arrow F2 is located inside a sub-phase (Fig. 1).

In Function F3, the teacher redirects or refocuses the students' thinking. She helps students to rethink their issue and move to the next sub-phase (arrow F3 between sub-phases in Fig. 1). It is essential that the on-the-fly assessment moves students to a new direction when compared to what they were doing.

In Function F4, the teacher reviews students' comprehension before or after an inquiry activity. Thus, the teacher may look forward if she is setting up an activity (arrow F4 before

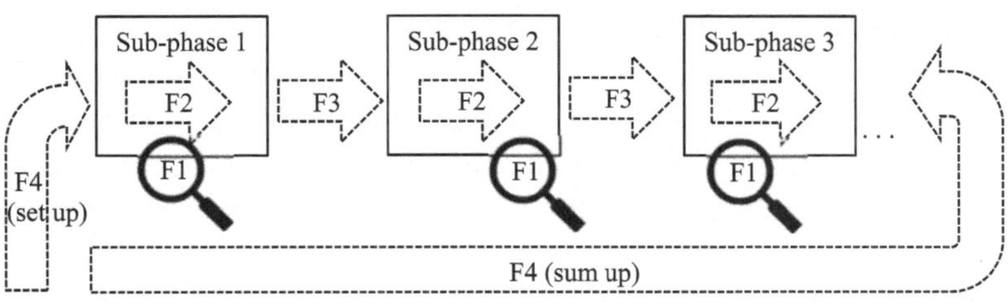

Figure 1. Functions of the on-the-fly assessment (F1–F4) when the teacher is supporting students along hypothetical sub-phases of their inquiry process. The arrows represent the teachers' aim to guide learning inside the sub-phases or to the next sub-phase and to set up or sum up learning. The magnifier refers to the teachers' aim to check or examine student thinking but not to change it.

Sub-phase 1 in Fig. 1). Alternatively, she may look backward if she is summing up what has happened (F4 after the last sub-phase). She may use information from many sources, which is illustrated as a long arrow.

The categorisation of forms and functions reveals substantial variation of the teachers' formative actions when examining only two classes. It is interesting that sometimes the forms of two episodes are the same, such as in Extracts 2 and 8 as well as in Extracts 4 and 5, but still the functions are different. On the other hand, we have the impression that there could be some relations between form and function. For example, all the episodes where the teachers corrected student thinking (F2a and F2b) were related to the collect by listening or observing and to the use by closed questions. However, our purpose is not to make any kind of generalisations and, with data only from the two lessons, it would not even be possible.

Many previous studies have presented the purposes and aims of informal formative assessment, albeit analysis usually has been done on the level of speaking turns. Many kinds of teacher moves have been reported. For example, teachers may give feedback, provide neutral or evaluative responses, ask for elaboration, ask for clarification, connect students' responses, promote exploration of students' own ideas and make connections to previous learning (Sezen-Barrie & Kelly, 2017; Ruiz-Primo & Furtak, 2006, 2007). Thus, possibly all the teacher moves we found from our data have been reported earlier. However, the novelty value of our research is in considering separately the forms and functions of on-the-fly and, especially, our attempt to link formative functions to the flow of student inquiry (see Fig. 1).

Differences in the forms and functions of on-the-fly assessment in the two lessons

Our third research question concerned the differences between the two lessons that can be identified by analysing the forms and functions of formative assessment. We found that there were plenty of differences in the functions between the two lessons, but fewer in the forms. The functions in Brown's class included more student-based guidance (F1c, F2b and F2c) whereas in Smith's class functions F1a and F2a indicate more teacher-centred guidance. It seems the teachers made different pedagogical choices how to guide students. Smith attempted to correct students' observations during the inquiry phase whereas Brown tried more to support students' own ideas although they were not necessarily correct. In other words, Brown's style of guiding students is more dialogic (Mortimer & Scott, 2003) or multi-voiced (Dysthe, 1996), increasing the presence of students' ideas in discussions. These differences can be seen also in the structures of the lessons: Smith used less time for the inquiry phase and proceeded faster in her guiding than Brown.

In addition, Brown's lesson included more of function F4 than Smith's lesson. This fits with our previous interpretation that Brown tried to support students' own ideas because then it is natural to aim for closure in the whole class discussion where students' results and ideas are discussed and the class together arrives at scientific explanations. In these discussions, Brown used her findings about students' incorrect ideas raised during the inquiry phase in order to guide students' thinking towards scientific concepts. Students' ideas were included and developed in the discussion, and thus, the discussion was dialogic (Mortimer & Scott, 2003).

In summary, although the two lessons were similar when comparing the forms of on-the-fly assessment, the differences in functions of on-the-fly assessment revealed differences in how teacher-centred or authoritative the assessment is.

Conclusions

This paper presents novel results of the forms and the functions of on-the-fly formative assessment in a physics inquiry. Especially the functions provide an informative view of the versatile actions teachers used to support student learning. In addition, we present a diverse view of the ways teachers may collect information from students, which has not been previously widely reported. This gives also more information about student-initiated assessment conversations that are less studied in previous research (Ruiz-Primo, 2011).

This study did not attempt to generalise any of its findings. In the future, the forms and functions as well as possible relations between them should be studied using a larger data set from more than two teachers. Then, it might be possible to make generalisations about the relations of forms and functions but also different characteristics of classes and teachers in relation to the observed on-the-fly formative assessment practices. However, most of the differences between the two lessons were noticed in functions of the on-the-fly assessment. This points to the importance of studying functions in future studies in assessment conversations, particularly when comparing lessons.

In this study, the student learning was not assessed using any external measure such as a test, and it would not even have been reasonable because of the short, one-lesson intervention. However, the functions demonstrate teacher moves that may help students to understand and to proceed in their inquiry, but the progress has to be interpreted from each situation. In future research, it would be possible to study in detail how students complete inquiry sub-phases or proceed from one sub-phase to another as an evidence of learning during the inquiry process.

Disclosure statement

No potential conflict of interest was reported by the authors.

Funding

This work was supported by the FP7 Science in Society [grant number 321428].

ORCID

Pasi Nieminen http://orcid.org/0000-0001-6019-1054
Markus Hähkiöniemi http://orcid.org/0000-0002-6337-5366
Jouni Viiri http://orcid.org/0000-0003-3353-6859

References

Assessment Reform Group. (2002). Assessment for learning: 10 principles. Retrieved from http://www.hkeaa.edu.hk/DocLibrary/SBA/HKDSE/Eng_DVD/doc/Afl_principles.pdf

Bell, B., & Cowie, B. (2001). The characteristics of formative assessment in science education. *Science Education, 85*(5), 536–553.

Bell, R. L., Smetana, L., & Binns, I. (2005). Simplifying inquiry instruction. *The Science Teacher, 72* (7), 30–33.

Black, P., & Wiliam, D. (2009). Developing the theory of formative assessment. *Educational Assessment, Evaluation and Accountability (Formerly: Journal of Personnel Evaluation in Education), 21*(1), 5–31.

Brookhart, S. M. (2015). *How to make decisions with different kinds of student assessment data.* Alexandria, VA: ASCD.

Chin, C. (2007). Teacher questioning in science classrooms: Approaches that stimulate productive thinking. *Journal of Research in Science Teaching, 44*(6), 815–843.

Chin, C., & Brown, D. E. (2002). Student-generated questions: A meaningful aspect of learning in science. *International Journal of Science Education, 24*(5), 521–549. doi:10.1080/09500690110095249.

Constantinou, C. P., & Papadouris, N. (2016). *Synthesis of factors determining the uptake of various assessment methods in different educational settings (Deliverable 5.11).* A report from the FP7 project assess inquiry in Science, Technology and Mathematics Education. Retrieved from https://assistme.ku.dk/pdf-uploads/D5.11.pdf

Corbin, J., & Strauss, A. (2008). *Basics of qualitative research: Techniques and procedures for developing grounded theory* (3rd ed.). Thousand Oaks, CA: Sage.

Correia, C. F., Nieminen, P., Serret, N., Hähkiöniemi, M., Viiri, J., & Harrison, C. (2016). Informal formative assessment in inquiry-based science lessons. In J. Lavonen, K. Juuti, J. Lampiselkä, A. Uitto, & K. Hahl (Eds.), *Electronic proceedings of the ESERA 2015 conference. Science education research: Engaging learners for a sustainable future, Part 11: Evaluation and assessment of student learning and development, co-ed. Jens Dolin & Per Kind* (pp. 1782–1791). Helsinki: University of Helsinki. ISBN 978-951-51-1541-6.

Dolin, J., Black, P., Harlen, W., & Tiberghien, A. (2018). Exploring relations between formative and summative assessment. In J. Dolin & R. Evans (Eds.), *Transforming assessment: Through an interplay between practice, research and policy* (pp. 109–140). Cham: Springer International Publishing.

Dysthe, O. (1996). The multivoiced classroom: Interactions of writing and classroom discourse. *Written Communication, 13*(3), 385–425. doi:10.1177/0741088396013003004

Furtak, E. M., Kiemer, K., Circi, R. K., Swanson, R., de León, V., Morrison, D., & Heredia, S. C. (2016). Teachers' formative assessment abilities and their relationship to student learning: Findings from a four-year intervention study. *Instructional Science, 44*(3), 267–291. doi:10.1007/s11251-016-9371-3

Furtak, E. M., Ruiz-Primo, M. A., & Bakeman, R. (2017). Exploring the utility of sequential analysis in studying informal formative assessment practices. *Educational Measurement: Issues and Practice, 36*(1), 28–38. doi:10.1111/emip.12143.

Goldberg, F. M., & McDermott, L. C. (1986). Student difficulties in understanding image formation by a plane mirror. *The Physics Teacher, 24*(8), 472–481. doi:10.1119/1.2342096

Harrison, C., Constantinou, C. P., Correia, C. F., Grangeat, M., Hähkiöniemi, M., Livitzis, M., … Viiri, J. (2018). Assessment on-the-fly: Promoting and collecting evidence of learning through dialogue. In J. Dolin & R. Evans (Eds.), *Transforming assessment: Through an interplay between practice, research and policy* (pp. 83–107). Contributions from Science Education Research, 4. Cham: Springer International Publishing. doi:10.1007/978-3-319-63248-3_4.

Hattie, J., & Timperley, H. (2007). The power of feedback. *Review of Educational Research, 77*(1), 81–112. doi:10.3102/003465430298487

Hobson, E. H. (1997). Introduction: Forms and functions of formative assessment. *The Clearing House: A Journal of Educational Strategies, Issues and Ideas, 71*(2), 68–70.

Lehesvuori, S., Viiri, J., Rasku-Puttonen, H., Moate, J., & Helaakoski, J. (2013). Visualizing communication structures in science classrooms: Tracing cumulativity in teacher-led whole class discussions. *Journal of Research in Science Teaching, 50*(8), 912–939. doi:10.1002/tea.21100.

Lobato, J., Clarke, D., & Ellis, A. B. (2005). Initiating and eliciting in teaching: A reformulation of telling. *Journal for Research in Mathematics Education, 36*(2), 101–136. doi:10.2307/30034827.

Louca, L., Zacharia, Z., & Tzialli, D. (2012). Identification, interpretation—evaluation, response: An alternative framework for analyzing teacher discourse in science. *International Journal of Science Education*, *34*(12), 1823–1856. doi:10.1080/09500693.2012.671971.

Mehan, H. (1979). *Learning lessons: Social organization in the classroom*. Cambridge, MA: Harvard University Press.

Miles, M. B., & Huberman, A. M. (1994). *Qualitative data analysis* (2nd ed.). Thousand Oaks, CA: Sage.

Mortimer, E. F., & Scott, P. (2003). *Meaning making in secondary science classrooms*. Buckingham: Open University Press.

Myhill, D., & Dunkin, F. (2005). Questioning learning. *Language and Education*, *19*(5), 415–427. doi:10.1080/09500780508668694

Nieminen, P., Correia, C., Hähkiöniemi, M., Serret, N., Viiri, J., & Harrison, C. (2016, April). *Formative assessment in inquiry-based science education using interactions on-the-fly*. Paper presented at the NARST Annual International conference, Baltimore, MD. https://doi.org/10.13140/rg.2.1.1162.5363

Nieminen, P., Hähkiöniemi, M., Leskinen, J., & Viiri, J. (2016). Four kinds of formative assessment discussions in inquiry-based physics and mathematics teaching. In H. Sifverberg & P. Hästö (Eds.), *Annual symposium of the Finnish mathematics and science education research association 2015* (pp. 100–110). Turku: University of Turku. ISBN 978-952-93-8233-0.

Oliveira, A. W. (2010). Improving teacher questioning in science inquiry discussions through professional development. *Journal of Research in Science Teaching*, *47*(4), 422–453. doi:10.1002/tea.20345

Pedaste, M., Mäeots, M., Siiman, L., De Jong, T., Van Riesen, S., Kamp, E., ... Tsourlidaki, E. (2015). Phases of inquiry-based learning: Definitions and the inquiry cycle. *Educational research review*, *14*. https://doi.org/10.1016/j.edurev.2015.02.003

Rocard, M., Csermely, P., Jorde, D., Lenzen, D., Walberg-Henriksson, H., & Hemmo, V. (2007). *Science education now: A renewed pedagogy for the future of Europe*. Brussels: European Commission.

Rönnebeck, S., Nielsen, J. A., Olley, C., Ropohl, M., & Stables, K. (2018). The teaching and assessment of inquiry competences. In J. Dolin & R. Evans (Eds.), *Transforming assessment. Contributions from science education research* (pp. 27–52). Contributions from Science Education Research, 4. Cham: Springer International Publishing.

Ruiz-Primo, M. A. (2011). Informal formative assessment: The role of instructional dialogues in assessing students' learning. *Studies in Educational Evaluation*, *37*(1), 15–24. doi:10.1016/j.stueduc.2011.04.003.

Ruiz-Primo, M. A., & Furtak, E. M. (2006). Informal formative assessment and scientific inquiry: Exploring teachers' practices and student learning. *Educational Assessment*, *11*(3–4), 205–235.

Ruiz-Primo, M. A., & Furtak, E. M. (2007). Exploring teachers' informal formative assessment practices and students' understanding in the context of scientific inquiry. *Journal of Research in Science Teaching*, *44*(1), 57–84. doi:10.1002/tea.20163

Ruiz-Primo, M. A., & Kroog, H. (2017, April). *Variations of formative assessment practices across instructional tasks: The case of informal formative assessment practices during whole class discussion*. Paper presented at the meeting of the American educational research association. San Antonio, TX.

Saxe, G. B., Gearhart, M., Franke, M. L., Howard, S., & Crockett, M. (1999). Teachers' shifting assessment practices in the context of educational reform in mathematics. *Teaching and Teacher Education*, *15*(1), 85–105.

Scott, P. H., Mortimer, E. F., & Aguiar, O. G. (2006). The tension between authoritative and dialogic discourse: A fundamental characteristic of meaning making interactions in high school science lessons. *Science Education*, *90*(4), 605–631. doi:10.1002/sce.20131

Sezen-Barrie, A., & Kelly, G. J. (2017). From the teacher's eyes: Facilitating teachers noticings on informal formative assessments (IFAs) and exploring the challenges to effective implementation. *International Journal of Science Education*, *39*(2), 181–212. doi:10.1080/09500693.2016.1274921.

Shavelson, R. L., Young, D. B., Ayala, C. C., Brandon, P. R., Furtak, E. M., Ruiz-Primo, M. A., ... Yin, Y. (2008). On the impact of curriculum-embedded formative assessment on learning: A collaboration between curriculum and assessment developers. *Applied Measurement in Education*, *21*, 295–314. doi:10.1080/08957340802347647.

van Zee, E., & Minstrell, J. (1997). Using questioning to guide student thinking. *Journal of the Learning Sciences*, *6*(2), 227–269. doi:10.1207/s15327809jls0602_3

Westgate, D., & Hughes, M. (1997). Identifying 'quality' in classroom talk: An enduring research task. *Language and Education*, *11*(2), 125–139. doi:10.1080/09500789708666723

White, R., & Gunstone, R. (1992). *Probing understanding*. New York, NY: Routledge.

Wiliam, D., & Black, P. (1996). Meanings and consequences: A basis for distinguishing formative and summative functions of assessment? *British Educational Research Journal*, *22*(5), 537–548. doi:10.1080/0141192960220502

Appendix: An example of a student worksheet

Task 2. Two persons are sitting in front of a plane mirror. The object is not perpendicularly in front of the mirror. The mirror is covered with a cloth.

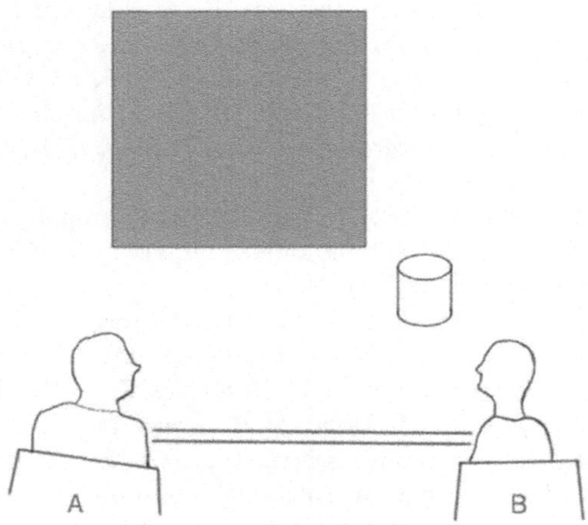

PREDICTION. With your peer, think about the following questions without the mirror. If the cloth is taken away,

a) does the person A see the object from the mirror? Why?
b) Does the person B see the object from the mirror? Why?

INQUIRY. Investigate the previous questions with a mirror. Mark with x the location of the image from the perspective of A and B.

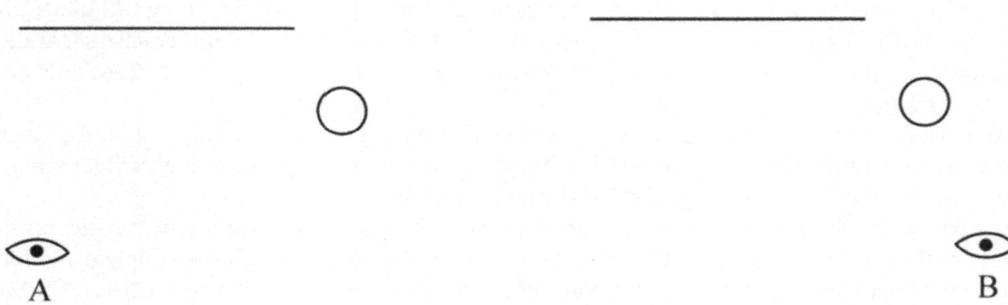

CONCLUSION. Did your prediction and observation differ? How does the person's position affect the seeing of the mirror image?

French teachers' informal formative assessment in the context of inquiry-based learning

Elie Rached and Michel Grangeat

ABSTRACT

Discursive formative assessment (FA) practices during instruction have grown in importance in the education literature as a way of improving teaching and learning. This study involves three years of design-based research as part of an educational project. We examine lower secondary school teachers' qualitative Informal Formative assessment (IFA) practices during inquiry-based Science, Mathematics and Technology-Engineering (STEM) lessons. In-service teachers ($N = 9$) were filmed during a whole-classroom discussion session with respect to students' alternative solutions or predictions in the context of problem-solving or investigating a natural phenomenon. The analysed discussions in the project's third year lasted 15–30 minutes. The ESRU (Eliciting, Student response, Recognising and Using phases) model is adapted to analyse IFA practices, while also considering the role played by students. Inquiry dimensions (e.g. conceptual and epistemic) were used to examine the inquiries given priority in these discussions. The results show frequent (>30%) ESRU complete-cycle implementations by most teachers, which is associated with effective IFA practices. Frequent complete-cycle implementation led to teachers eliciting the same students at different moments of the session with longer consecutive cycles. The dimensions the teachers implement are more epistemic than conceptual. Moreover, students took charge of several ESRU phases.

Introduction

Examining the nature of talk that occurs in classrooms is vital for shedding light on teaching-learning practices, especially conversations that allow insight to the formative assessment practices underway during instruction (Duschl, 2008; Ruiz-Primo & Furtak, 2007).

The literature indicates that observations of formative assessment (FA) in the daily life of classrooms are valuable (Allal & Mottier Lopez, 2005; Cizek, 2010). A particular form of classroom discursive process called 'assessment conversations' is distinguished, which is associated with an effective Informal Formative Assessment (IFA), a discursive and unplanned form of classroom FA (Ruiz-Primo & Furtak, 2007). However, many challenges remain in relation to the development of these practices and of the frameworks that describe FA in inquiry-based classroom contexts (Duschl & Gitomer, 1997; Ruiz-Primo & Furtak, 2007).

This study examines lower secondary Science, Technology/Engineering[1] and Mathematics (STEM) teachers who daily implement IFA practices in inquiry-based education (IBE) courses. Teacher–student(s) and student(s)–student(s) IFA conversations occurring during a whole-classroom discussion session are analysed.

The first section presents the research background, including a description of the ESRU model for analysing IFA in the context of IBE. After setting the research questions, we describe the methodology for collecting and analysing the classroom discussions before presenting the findings. Finally, the research results and prospects are discussed.

Research background

From inquiry-based STEM education to classroom discourse

Education policies aim to engage students in scientific inquiry and engineering design (Hazelkorn et al., 2015; NRC, 2015). The science education literature indicates that inquiry in science education is a term used to renew science teaching-learning and assessment (Abd-El-Khalick et al., 2004; Duschl, 2008). It is also linked to the 'design process' concept in technology education and to the Problem Solving (PS) concept in mathematics education (Artigue & Blomhøj, 2013; Bernholt et al., 2013). In France, guided by teaching strategies which emphasise scientific investigation, these approaches include more student-centred learning with a focus on teaching-learning processes (Lebeaume, 2011). Nonetheless, the French IBE curriculum stresses both analogies that embrace different disciplines in Sciences/mathematics/technology and the specific features of disciplines (MEN, 2008).

Parallel to this, STEM education is expected to link science with other subjects and disciplines for an 'integrated STEM' education as opposed to education with more STEM-isolated disciplines (Brown, 2012; Hazelkorn et al., 2015). STEM education is also expected to add to our understanding and knowledge of scientific principles (Hazelkorn et al., 2015). However, which 'form(s) of integration' or 'inter-disciplinarity' are to be introduced in STEM education is a major challenge in the literature (Wong & Dillon, 2019). Moreover, framing the 'common' aspects of a STEM inquiry(ies) across domains remains debated and strongly linked to science education (Rudolph, 2000). Still, some researchers identify common IBE aspects in literature that may be associated with an inquiry-based education in STEM (Artigue & Blomhøj, 2013; Bernholt et al., 2013) such as: diagnosing a problem/phenomenon; formulating a hypothesis; planning investigations/designs; constructing and using models; constructing and criticising arguments; debating with peers/communicating etc.

Teaching and learning approaches in STEM share goals/characteristics with inquiry-based science education, albeit they are differently labelled (Harisson, Grangeat and Dolin, this issue).

In a sociocultural perspective, learning is mainly mediated through talk and the words used in social exchanges provide the tools needed for individual thinking (Mercer et al., 2004; Mortimer & Scott, 2003). Science education is portrayed as a discursive process where novices (students) are invested into representing and understanding phenomena by using language and other representational means, so as to enable persons (e.g. the teacher and students) who are more expert in the field to support others in making sense of the scientific view introduced in the classroom (Mercer et al., 2004; Mortimer & Scott, 2003). Hence, if we expect students to engage in meaningful classroom science learning they should engage in the dialogic process of exploring and working on ideas (Scott et al., 2006).

A science education approach of this nature requires that conditions be created in which students can engage in (rich) argumentation 'to explore critically the coordination of evidence and theory that support or refute an explanatory conclusion, model or prediction' (Duschl & Osborne, 2002, p. 44), a major feature of scientific inquiry. Moreover, argumentation allows the possibility of making students' thinking 'visible' through presentations and discussions and so offering teachers' FA opportunities regarding their students (or students regarding their peers) (Black & Wiliam, 2003; Osborne et al., 2004).

Accordingly, examining the nature of talk (and other communication modes) is vital for shedding light on the teaching-learning practices that occur in classrooms where, in particular, examining (teacher–student(s) and student(s)–student(s)) classroom assessment conversations allows insights into IFA practices (Ruiz-Primo & Furtak, 2007).

Formative assessment and learning

Many educators rely on assessment to improve learning (and teaching), sometimes called 'assessment for learning' or 'formative assessment' (FA) (Allal & Mottier Lopez, 2005; Black & Wiliam, 2003).

Researchers state it is important to use evidence of student achievement to decide on potentially adapting the instruction, and emphasise that FA assists in regulating teaching and learning (Allal & Mottier Lopez, 2005).

Implementing FA in the classroom is generally described by asking three questions (Ruiz-Primo et al., 2010): Where should the teaching-learning processes lead to? Where are they now? How to get there?

Researchers acknowledge three forms of FA: *informal*, *planned* and *formal* (Ruiz-Primo et al., 2010). In particular, Ruiz-Primo and Furtak (2007, p. 59) argue that IFA practices occur when the timeframe for interpreting and acting is more immediate than other modalities of FA and imply that 'unplanned' FA takes place during instruction 'in any student–teacher interaction at the whole-class, small-group, or one-on-one level'. These authors distinguish the specific use of dialogue in the form of assessment processes that encourage classroom inquiry. Called 'assessment conversations', these dialogues 'permit teachers to recognize students' conceptions, mental models, strategies, language use, or communication skills, and allow them to use this information to guide instruction' (Ruiz-Primo & Furtak, 2007, p. 60). These assessment conversations share three characteristics of effective IFA (Eliciting, Recognising and Using).

Consequently, observing FA in the daily life of the classroom is valuable (Allal & Mottier Lopez, 2005; Cizek, 2010). Moreover, questions continue about how to support teachers to develop these forms of conversations in an inquiry-based science learning context (Duschl & Gitomer, 1997; Ruiz-Primo & Furtak, 2007). Another challenge is to develop a framework for describing and analysing IFA practices in these contexts (Ruiz-Primo & Furtak, 2007). Moreover, research stresses the need to consider students' involvement in FA practices (Allal & Mottier Lopez, 2005; Black & Wiliam, 2009).

Gaining access to IFA practices in the context of IBE

The ESRU model for analysing IFA

Establishing teacher–student classroom discussions as a major element within the day-to-day course of the classroom, some researchers have developed an analytical model to gain access to the IFA taking place during inquiry-based science units. Authors describe the 'assessment conversations' associated with effective IFA as combinations of complete ESRU cycles where

> *the teacher Elicits a question* [allowing the students to share and make visible or explicit their understanding as fully as possible]*; the Student responds; the teacher Recognises the response* [judging differences between students' responses and learning goals and making explicit the critical dimensions relevant for improving their learning]*; and then Uses the information collected to support the students' learning* [acting with respect of students' responses to help them 'progress' towards their learning goals]. (Ruiz-Primo & Furtak, 2007, p. 61)

A cycle usually lasts a few seconds.

The authors distinguish ESRU from earlier models (e.g. the Initiate, Respond and Evaluate (IRE) or Feedback (IRF) models) dedicated to analysing classroom discourse and differentiate it from other forms of conversations in everyday life. Some IRE/F cycles imply 'inauthentic teachers' questions' where the answer is already known by the teacher, who engrosses the students in 'procedural' rather than 'authentic' engagement (Cazden, 2001; Mehan, 1979). For instance, these IRF/E cycles continue until the students 'guess' the right answers, ending up with the teacher searching for the 'one correct' answer, while some good 'possible' answers might be missed and other answers, considered correct by the teacher, do not allow to confirm if the student's answer comes from mastery of the concept or from the conversation (Mehan, 1979). On the contrary, 'assessment conversations' require teachers to be mediators of student learning rather than simple providers or evaluators of acceptable answers (Ruiz-Primo & Furtak, 2007). Hence, the 'Eliciting' initiating questions in the ESRU model would provide information about the evolving status of students' conceptions and understanding. 'Recognising' the students' responses should indicate that the students' contribution has been heard and accepted as part of the ongoing classroom narrative and gives students the opportunity to evaluate the correctness of the teacher's interpretation of the student's contribution. Through the 'Use' aspect of the discussion, the teacher provides students with specific information on actions they can take to accomplish the learning goals (e.g. ask another question that challenges the students' thinking, 'Compares/contrasts the students' responses', 'Promotes debating/discussion', 'Promotes idea exploration' and 'Promotes thinking' (asking 'Why/how?)' etc.).

Analysing inquiry dimensions

Ruiz-Primo and Furtak (2007) analysed what type of inquiry is promoted during IFA and ESRU cycles. Their model involves three main inquiry dimensions: epistemological, conceptual and social.

- The epistemological framework includes both the abilities involved in the processes of science (e.g. observing, hypothesising and experimenting, using evidence, logic and knowledge to construct explanations) and knowledge of the rules and criteria to judge what counts as scientific inquiry (e.g. experiments, hypotheses, explanations).

- The conceptual structure includes a deep understanding of the concepts and principles as part of larger scientific conceptual schemes.
- The social processes dimension focuses on how knowledge is communicated, represented and argued. This dimension refers to frameworks involved in students' scientific communications needed to engage in scientific inquiry and can be oral, written or pictorial.

Hence, argumentation is considered to be part of social processes of engaging in scientific inquiry and an integrated element of the ESRU model. Moreover, researchers consider that any assessment conversation is in its nature a social interaction between the teacher and students and it is difficult to distinguish it from the epistemic and conceptual domains and the social dimension is accordingly integrated into the conceptual and epistemic dimensions.

Researchers also highlight the inquiry dimensions seen in the Eliciting (E) phases of ESRU cycles. As the first element of a cycle, Eliciting helps determine which type of inquiry dimension is 'intended' for each cycle.

Examples of eliciting the epistemic dimension are as follows:

Asking the students to: 'formulate an explanation', 'provide evidence and examples', 'relate the evidence to the explanation', 'Evaluate the quality of the evidence', 'Compare/contrast the observations/data/procedures', 'Interpret the information/data/patterns', 'Compare/contrast other ideas' and 'promote debate'.

Examples of eliciting the conceptual dimension are: 'Apply/relate/compare/contrast concepts' and 'Check their comprehension'.

Research questions

Inquiry-Based Education in STEM and IFA challenges teachers' practices. This article focuses on lower secondary teachers' IFA and STEM-inquiry dimensions implemented in a whole-classroom-discussion session as part of inquiry-based lessons.

The research questions are:

(1) Which IFA practices described by the ESRU model do STEM teachers implement in inquiry-based classroom lessons?
(2) What is the role of students in these IFA processes?
(3) What is the nature of the inquiries (namely, the conceptual or epistemic dimension) promoted in these IFA processes?

Methodology

Our approach to conducting the research is based on 'Design Experiments' (Cobb et al., 2003). Design experiments are situated and entail both the 'engineering' of particular forms of learning and the means to support them ('learning ecology'), along with systematically studying such forms of learning.

Participants

The study took place in two low-income lower secondary schools in a multi-ethnic area on the outskirts of Grenoble (France). The group of teachers whose practices have been analysed were participating in a three years educational research project. Teachers had more than five years' teaching experience (except for the Mat-II mathematics teacher, a beginner) and had volunteered to cooperate with teacher-educators and researchers on developing classroom FA in IBE. The group comprised three mathematics (Grade 6), three physics (Grade 7), one biology (Grade 6) and two technology (Grade 8) in-service teachers (five females and four males). All of the teachers have higher education (bachelor or higher) degree concerning the topic they teach.

Instructional context

The three years project emphasised different aspects entailed in changing teacher practices (for more details, see Grangeat and Lepareur in this issue). Three seminars per year were created in order to explore the shared problem (integrating STEM inquiry-based education and FA practices in the classroom). The project's first year was spent elaborating on a common understanding of IBE and FA and designing and testing lessons and FA tools for each topic. The second-year involved refining the products of the first year. Teachers were asked to design and implement a common FA tool (i.e. a rubric) and discuss the ways it had been used in their classroom.

In the third year, teachers, regrouped in teams by the topic to be taught, were asked to design a lesson plan that incorporates common aspects of inquiry-based education in STEM (e.g. diagnosing a problem/phenomenon; formulating (choosing) the hypothesi (e)s/prediction(s), solution(s)/conjecture(s) or artefact(s)/design(s); planning investigations/design; constructing and criticising arguments or explanations/using evidence; debating with peers/communicating) and FA tools.

The teachers were also encouraged to discuss classroom implementation of the lesson plan. Each trial included three teachers, with each teacher implementing in their classroom one of four topics in STEM. Hence, after each trial, discussions with the research team were organised, and then a new teacher volunteered to teach the updated lesson plan in the next trial. This iterative process involved a series of three trials, one per semester. The teachers made no major changes to these lesson plans between the trials, except for one technology (Tec-II) teacher who added a simulation activity in the second session. Extracts of the video recorded classroom practices and interviews with teachers were discussed collectively.

The data analysed in this article relate to the experiment conducted in the third year.

Moreover, the researchers stimulated the teachers to address argumentation in their lesson by encouraging students (individually and/or in teams) to practise argumentation (in written and oral form) by presenting their claim and the reasons in support of their claim; confronting students with alternative predictions and solutions, exposing them to other students' arguments and counter-arguments and, in some cases, to think how to respond to those counter-arguments.

Lesson plan

A lesson plan consists of four sessions. In the first session, teachers present students with the topic to be investigated, the lesson aims and tasks. The teaching topics were: Physics (PC) – state of matter change (heated water (till boiling) temperature evolution), Biology (Bio) – germination of grains (environmental factors impacting germination),

Technology/engineering (Tec) – designing/constructing a paper bridge, and Mathematics (Mat) – fractions (finding the size of a garden by using a particular unit).

The teachers also prepare a common questionnaire regarding the teaching goals and students' predictions/solutions to the issue being investigated and hand it out to the students at the start of the lesson. In addition, written FA feedback on (a sample of) the students' answers regarding the questionnaire is given by the teachers. In the second session, the students (individually or in teams) make prediction(s)/hypotheses (science), solution (s)/conjectures (mathematics) or artefact(s)/design (technology) regarding the phenomenon/problem and the reason(s) in support, for a presentation to the whole classroom. Teachers go around the student teams and encourage them to clarify and justify their hypotheses/predictions/solutions/designs. However, teachers do not intervene with respect to the 'correctness' of the students' answers. In biology and technology units, students also launched respectively their experiment on the environmental factors of 'grain germination' and design a bridge mock-up.

In the third session, the student teams present their respective predictions/solutions. A whole-classroom discussion is organised around these presentations, during which the students deliberate other teams'/students' predictions/solutions and consider possible counter-arguments.

In the fourth session, students test (experiment) their predictions/solutions, develop the conclusions of the finished work and/or complete exercises on the topic being taught.

At the end of each session, the students complete a self-assessment grid regarding the teaching goals.

On different occasions, the teachers make an IFA of the students: as they complete a self-assessment grid, work in teams to prepare their respective predictions/solutions and arguments (e.g. session 2-3) and conduct their experiments (e.g. session 4).

Data collection

Three sessions per teacher were recorded by a camera positioned at the back of the classroom. This study analyses session 3, whole-classroom discussion. All session 3 video recordings were transcribed.

The whole-classroom discussion in physics, technology and biology lasted less than 30 minutes. However, the mathematics discussions lasted more than one hour. To compare the four topics, three extracts of mathematics teachers were selected to cover session 3. The extracts included: three students' proposed solutions discussions, at the beginning, in the middle and at the end of the session. The three proposed solutions were the same for all mathematics teachers. Consequently, the analysed episodes on average last 18 minutes, ranging from 10 to 30 with a median of 20 (see Table 4).

Data analysis

We employ micro and intermediate levels of analysis (Tiberghien et al., 2007) to examine the IFA practices that were being promoted during the whole-classroom discussion. At the micro level, we also analyse the nature of inquiries. Nevertheless, our attribution of the levels relates more to the nature of teacher–student(s) interactions.

The micro level focuses on analysing one sequence of either a complete Eliciting, Response, Recognising and Using (ESRU) cycle, namely effective IFA practices, or uncomplete (ES/ESR) cycles, namely ineffective IFA practices, as promoted during the teacher–students and student(s)–student(s) interactions. We looked at the transcriptions of the 'assessment conversations' where the content at stake related to the inquiry lesson topic and the teacher was not the only speaker. We assigned a code (e.g. E, S, R or U) to each intervention according to its role in the ESRU model.

Intervention elements like 'excellent' or 'yes' not rooted in the assessment of inquiry were coded as no (inquiry) assessment interventions (NI). Further, some of the E, R and U elements of a cycle might be implicit. An implicit phase requires the researcher to make an interpretation after taking the whole context of the discussion into account.

These were coded within brackets (e.g. [E], [R] and [U]). Moreover, different numbers were used to separate the different cycle sequences (e.g. E1, S1; [E2], S2, R2, U2) (for example, see Tables 1 and 2). Finally, we coded the eliciting (E) part of a cycle by stating whether it is an inquiry of the 'epistemic dimension' or 'conceptual dimension'.

We also analysed the teacher–student(s) interactions at the intermediate level. The intermediate level is based on the work of Ruiz-Primo and Furtak (2007) work to establish the nature of the teacher–student(s) interaction in a session by determining if teachers implement any consecutive cycles with the same student (≥2 complete or uncomplete cycles) and the number of times such consecutive cycles are observed with the same student. Nevertheless, we also consider whether these consecutive cycles included interactions with other students (C) and if these successive cycles contained both epistemic and conceptual dimensions (M).

Our first example is teacher (PC-I) with regard to the physics topic of the evolution of water temperature during a state of matter change. In this example, the teacher seeks to establish what type of understanding underpins a student's choice of a graph and explanation of an associated phenomenon. The teacher completes five consecutive cycles with the same student (Margot) to investigate the student's ideas regarding this phenomenon.

On an intermediate level, from a sample of five teachers who represent different disciplines, trials and ESRU frequency (including the two teachers with the highest share of complete ESRU and the two teachers with the lowest one), we analysed teachers' IFA practices.

Table 1 gives examples of four complete ESRU cycles and one incomplete ESR cycle. The implicit parts, eliciting [E2] and recognising [R2], are noticeable. Inquiry dimensions are epistemic at 1, 4, 6, 12 and 14 T.

Table 2 presents an example of incomplete (ESR) and a complete ESRU cycle. The inquiry dimensions are epistemic at 1T1 and 3T1.

ESRU model adjustments

In this study, adjustments were made to the ESRU model considering the students' roles in elaborating certain ESRU cycle elements/phases, IFA non-verbal parts, and the complexity of the conversations.

ESRU adjustments regarding the students' role in IFA. Any effective classroom practice is, by nature, a mutual action engaged in by the teacher and the students (Tiberghien et al., 2007). Hence, during 'assessment conversations' particular students' interventions may be

Table 1. IFA implementation with complete ESRU cycles (physics teacher PC-I).

Turn taking	Transcription	Code
1 T (teacher)	*Those who chose the first proposition are ready to justify their choice. Margot we're listening to you.*	E1: Teacher asks student to Formulate explanation <u>*Epistemic dimension (ED)*</u>
2Margot (Student)	*I think it is the first proposition because the increase in water keeps the same rhythm. The increase is bit by bit. We didn't change the temperature of the heater. Even if we don't change the temperature it keeps increasing.*	S1 (Student answers)
3T	*So you think that it is the temperature that is increasing, ok*	R1: Teacher revoices students' words (summarises what student said) or repeats/paraphrases students words
4T	*So why the temperature keeps rising like that?*	U1: Teacher Promotes students' thinking by asking them to elaborate their responses (why/ how?)
–	–	[E2] [implicit part of the ESRU cycle]: teacher asks student to interpret information, data, patterns <u>*ED*</u>
5Margot	*It does stop at a moment*	S2 (Student answers)
6T	*When does it stop?*	[R2: implicit recognising of students' answer] U2: Teacher promotes exploration of students' own ideas; [E3] <u>*ED*</u>
7T	*Wait // I can't hear very well // pupils you are a bit troubling*	No inquiry (NI)/ class management
8Margot	*Water is going to be transformed into steam*	S3
9T	*So you think that when water is going to become steam the temperature will not rise*	R3: Teacher revoices students words
10T	*The others maybe you have things to say. I can help to conduct the debate but you should ask questions.*	NI
11T	*So you think that when water will become steam the temperature will stop rising*	R3: Teacher revoices
12T	*What temperature are you talking about?*	U3: Teacher promotes exploration of students' own ideas <u>*[E4] ED*</u>
13Margot	*of liquid water*	S4
14T	*Liquid water or steam water?*	[R4] U4: Teacher promotes exploration of students' own ideas [E5] <u>*ED*</u>
15Margot	*liquid water*	S5
16T	Liquid water ok?	R5: Teacher repeats

associated with one of the eliciting, recognising and/or using phases of the ESRU model and thus part for IFA processes. Since the ESRU model does not address this issue, we made adjustments by incorporating the students' interventions responsible for 'Eliciting' other students', 'Recognising' and even 'Using responses'. These phases were marked respectively as 'E', 'R' and 'U'.

The example is given in Table 3 illustrates how student's interventions (from group 1) at 4GR1 and 8GR1 contributed to the respective phases [R'1] and U'1 of complete ESRU cycle 1 (E1S1[R'1]U'1).

ESRU adjustments considering the classroom discussion's complexity and non-verbal interventions. In their review of how norms of scientific argumentation are established in the classroom, researchers note the limitations of the analytical models used in the science education literature to examine argument in a discourse (Driver et al., 2000). In particular, they state that in the natural flow of conversations points are not necessarily

Table 2. Example of IFA implementation by a Mathematics teacher (Mat-II).

Turn taking	Transcription	Code
1T1 (teacher)	*So you have some doubts? Who wants to give me a reason for which he voted true? Marie*	E1: Teacher asks students to formulate (scientific) explanation; *ED*
2Mari (student)	*I say it's true because I've tried two different ways and with both ways I found the same results.*	S
3T1	*You tried two different ways.*	R1: Teacher Repeats/paraphrases students' words
	I need another argument because you may be wrong twice.	U1: Teacher Promotes exploration of students' own ideas
	Yanis.	E2: Teacher asks students to Formulate an explanation; *ED*
4Yan (student)	*Because if we do, if the unit makes 12, 12 squares, we can take 9 times 12 squares and then 10 lefts*	S2
5T1	*We could take nine time twelve squares? Ehh* [the teacher writes on the board the answer given by the student.]	R2: Teacher Repeats

developed successively and reference must be made across extensive text parts to identify the features of the argument. Further, not all points are made with language but with gestures (e.g. pointing at objects), illustrations and graphics, especially in science.

We, therefore, adjusted the ESRU model to include the following aspects: non-verbal features attributed to one of the E, S, R or U cycle elements, as well as interventions by the teacher or students associated with different cycles.

As mentioned, in the example shown in Table 3 student intervention 8GR1 is part of the incomplete ESR cycle 2 and complete ESRU cycle 3 including an implicit [E3] part of cycle 2. In addition, the student's intervention at 11Dri contributed to both eliciting

Table 3. Example of IFA implementation by a technology teacher (Tec-I).

Turn taking	Transcription	Code
1T2 (teacher)	*Wait two seconds please*	NI
2T2	*Group 3.* [Teacher asks Group 3 to evaluate group 1 artefact's]	E1: Teacher asks students to: compare/contrast others' ideas; *ED*
3Dri (student from group 3)	*It's falling!* [Student pointing his finger to bridge artefact parts collapsing]	S1
4GR1 (student from group 1)	*Yeah, later we have just to reinforce here.* [Pointing to bridge parts that need consolidation]	[R'1] [implicit recognising] U'1: **Student** compares/contrasts students' responses to acknowledge and discuss alternative explanations conceptions
5T2	*Hey you listen. It doesn't seem that you understand. It's the third time I intervene.*	NI
6T2	*Dri, you were saying*	E2: Teacher contrasts others' idea; *ED*
7Dri	[it's not] *solid*	S2
8GR1	*and each time we try to make* [inaudible, adjustments] *we can't make it.*	R'2**: Student** Clarifies/Elaborates based on students' responses [E'3]
9Dri	*you need to do it*	S3
10T2	*They will* [do it] *eventually*	R3: Teacher Clarifies/Elaborates based on students' responses
11Dri	*They don't have time*	U'3: **Student** provides descriptive feedback [E'4] [S4]
12T2	*They don't have time.* *Ok. Other questions? Other critics? Constructive ones' for sure.*	R4: Repeats E5: Compare/contrast others' ideas

phases of the complete ESRU cycle 3 [E'3]S3R3U'3 and incomplete ESR cycle 4 [E'4][S4]R'4.

Moreover, the non-verbal features considered in the transcript (i.e. pointing at a graph) complemented the ESRU model and conceptual/epistemic dimensions. Hence, non-verbal interventions were coded as part of one of the phases (E, S, R or U) of the cycles as if they formed part of the assessment conversations, or as NI otherwise. For example, as shown in Table 3, at 3Dri and 4Gr1, the student pointing their finger to indicate that parts of the bridge are falling apart is analysed and integrated as part of the student response S1 in the E1S1[R'1]U'1 cycle.

Table 3 presents two complete ESRU cycles and two incomplete ESR cycles. The inquiry dimensions are epistemic (at 2T2 and 6T2).

STEM-Inquiry analyses

Researchers have examined scientific inquiry conducted in the context of physics teaching dedicated to IFA practices, while establishing the inquiry's epistemological, conceptual and social dimensions (Ruiz-Primo & Furtak, 2007). Nevertheless, these dimensions may be traced back to the broader inquiry approaches found in STEM education (Duschl, 2008).

However, the subcategories developed by the authors to analyse scientific inquiry were conducted in the context of physical science (e.g. Properties of Matter) teaching. Conversely, our data include lessons from different subject areas in STEM.

To reduce any gaps arising from use of the authors' inquiry dimensions with respect to STEM lessons, our analyses consequently focused on the main categories of epistemic and conceptual dimensions which might embrace broader inquiry dimensions in STEM and hence be treated as a general domain (and not domain-specific for a particular discipline in STEM) (Bernholt et al., 2013; Duschl, 2008; Harlen et al., 2012).

Validity and reliability of the qualitative analyses

To ensure the validity and reliability of the qualitative analyses, we proceeded as follows. In the initial phase of the analysis, the first author analysed a sample (around 20%) of data. The same sample was then independently analysed by the second author, after which the two authors compared their results and discussed any differences until agreement was reached. In the second phase, part of the sample was analysed by two more researchers from another partner-country involved in the research project and working on the same IFA practices. Then, any differences that emerged were discussed by the partners until a final consensus was achieved on 75% of the data analysed. In the third round, the first author completed a full analysis of all remaining data.

Results

To respond to research questions 1 and 2, we provide the frequency of each type of 'cycle' observed in all the assessment conversations in session 3, a whole-classroom discussion. To answer research question 3, we identified the eliciting phase (E) for both complete and incomplete IFA (ESRUs) cycles in the epistemic and conceptual dimensions.

Table 4. IFA results across teachers; frequency of ESRU, ESR and ES complete and incomplete cycles.

Trial	Teacher code/ minutes being analysed (total session 3 duration)	ESRU cycles frequency (%) Teacher	Student	ESR cycles frequency (%) Teacher	Student	ES cycles frequency (%)	No assessment inquiry interventions frequency (%)	Total cycles
I	Mat-I/ 30(1h40′)	39 (41.0%)	–	31 (32.6%)	–	13 (13.6%)	12 (12.6%)	95
	PC-I/ 20(26′)	19 (50%)	–	11 (28.9%)	–	4 (10.5%)	4 (10.5%)	38
	Tec-I/22(22′)	16 (25.8%)	5 (8.1%)	14 (22.6%)	1 (1.61%)	5 (8%)	21 (33.8%)	62
II	Mat-II/23(55′)	20 (16.1%)	1 (0.8%)	59 (47.6%)	1 (0.8%)	11 (8.8%)	32 (25.4%)	124
	PC-II/10(14′)	10 (20.7%)	6 (13%)	14 (30.43%)	–	4 (8.6%)	12 (26.0%)	46
	Tec-II/11(15′)	16 (31.4%)	1 (2%)	14 (27.4%)	–	7 (13.7%)	13 (25.4%)	51
III	Mat-III/14(1h10′)	30 (35.2%)	–	34 (40%)	–	8 (9.4%)	13 (15.2%)	85
	PC-III/13(15′)	9 (21.9%)	1 (2.4%)	25 (61%)	–	2 (4.9%)	4 (9.7%)	41
	Bio-III/22(24)	32 (32.6%)	1 (1%)	38 (38.7%)	–	4 (4%)	23 (23.4%)	98

Teachers' uptake of informal formative assessment

A complete ESRU cycle was the highest type of 'cycle' observed for five out of the nine teachers. In addition, complete cycles were totalised in more than 33.3% of all cycles by seven out of the nine teachers, representing frequent use of the ESRU complete cycle (ranging from 33.4% (Tec-II) to 50% (PC-I); see Table 4). However, complete cycles were observed less among two of the nine teachers, representing 24.3% (PC-III) and 16.9% (Mat–II) of total cycles, respectively.

Moreover, an incomplete ESR cycle was the highest type of 'cycle' observed among four of the nine teachers. Nevertheless, two of these teachers (Mat-III and Bio-III) also frequently used complete cycles (>33.3%).

Further, NI interventions were frequently used (33.8%) by one teacher (Tec-I) and comparable to the frequency of the other types of cycles. However, these did not limit Tec-I from frequently implementing complete cycles.

In addition, the frequency of each 'type' of cycle, for teachers teaching the same topic, may vary significantly. For example, PC-I accomplished the highest share of complete cycles (50%), while PC-III reached only half of that (24.3%). Nevertheless, 41.0% of cycles Mat-I implemented were complete, while the figure for Mat-II was 16.9%. It is worth noting that no topic- or discipline-specific trend was discernible in our data, although we note that a physics teacher (PC-III) made the greatest use of complete cycles among all the teachers and a mathematics teacher (Mat-I) made the lowest.

In addition, with six of the nine teachers, the students took part in elaborating many phases of the ESRU and ESR cycles. Yet, this was observed (<10%) relatively less than the teachers' efforts. Still, students' involvement represented a significant part (23.8% and 37.5%, respectively) of complete cycles for two of the teachers (Tec-I and PC-II).

Inquiry dimensions

All teachers in all trials and topics reveal far more epistemic frameworks – ranging from 84.2% (Tec-II) to 97.3% (PC-III) – than conceptual structures – ranging from 2.7% (PC-III) to 20.5% (PC-I) – in their inquiry dimensions (see Table 5).

Table 5. Frequency of inquiry dimensions during the Eliciting (E) phase of complete and incomplete ESRU cycles.

Trial	Teacher/discipline	Elicits epistemic frameworks frequency (percentage)		Elicits conceptual structures frequency (percentage)		Total Eliciting inquiry frequency
		Teacher	Student	Teacher	Student	
I	Mat-I	73 (87.9%)	–	10 (12.0%)	–	83
	PC-I	27 (79.4%)	–	7 (20.5%)	–	34
	Tec-I	38 (92.6%)	–	3 (7.3%)	–	41
II	Mat-II	83 (90.2%)	4 (4.3%)	5 (5.4%)	–	92
	PC-II	26 (76.5%)	5 (14.7%)	2 (5.9%)	1 (2.9%)	34
	Tec-II	32 (84.2%)	–	5 (13.1%)	1 (2.6%)	38
III	Mat-III	63 (87.5%)	–	9 (12.5%)	–	72
	PC-III	34 (91.9%)	2 (5.4%)	1 (2.7%)	–	37
	Bio-III	68 (90.6%)	–	7 (9.3%)	–	75

Once again, no differences were established between the topics being taught. Even here, within the same topic and same lesson, the nature of the dimensions elicited by the teachers can significantly vary. For example, PC-I implemented 20.5% of conceptual dimension (An example of the 'apply concept' strategy from PC-III: 'Ah so eh, for it to evaporate, it needs energy, or not?') while PC-III implemented merely 2.7%. In addition, Mat-I implemented 12.0% of conceptual dimension while Mat-II implemented only 5.4%.

Moreover, among four of the nine teachers, students participated in eliciting inquiry dimensions. Still, this participation was less frequent than teachers' eliciting efforts. However, for two teachers, PC-II and Tec-II, students were responsible for eliciting one-third and one over-sixth, respectively, of the conceptual inquiry dimension and one over-sixth of the epistemic ones' for PC-II.

We also examined the inquiry nature of the teachers' interventions in the Eliciting phase (see Table 6). We found that the epistemic dimension strategy of asking students to 'formulate an explanation' (category) was frequently observed among teachers who showed the highest complete cycle use (50% and 41% of total eliciting strategies by PC-I and Mat-I, respectively). Conversely, these strategies represented a mere 21% (Mat-II) and 29.3% (PC-III) among observed teachers demonstrating little use of a complete cycle. Moreover, the category 'Compare/contrast another's idea' eliciting the epistemic strategy was seen more with PC-I (15.8%), Mat-I (7.4%) and used considerably more by the teachers Tec-I (25.8%), but used less by PC-III (2.4%) and Mat-II (5.6%).

Nonetheless, eliciting the epistemic dimension category of 'Evaluate evidence quality', was the least observed teacher strategy and constituted 2.1% (Mat-I), 2.6% (PC-I), 4.8% (Mat-II) and 4.9% (PC-III), respectively. In addition, eliciting the conceptual dimension category of 'Apply/compare/contrast concepts' was the strategy most used by the teachers PC-I (15.8%) and Mat-I (10.5%) compared to Mat-II (1.6%) and PC-III (2.4%), where the latter employed no 'Check comprehension' strategies.

Intermediate analyses results

Teachers with a high share of complete ESRU cycle(see Table 7) tended to have longer exchanges (≥2 consecutive cycles) with the students, representing respectively 92.1% (PC-I), 80% (Mat-I) and 72.5% (Tec-I) of the total number of cycles than teachers with a low share of complete implementation, amounting to 51.2% (PC-III) and 28.2%

(Mat-II). Further, Mat-I and PC-I implemented more extended consecutive cycles (≥ 5 ES, ESR and/or ESRU iterations), representing for both 52.6% of total cycles, than teachers with a low frequency of complete cycles, namely 43% (PC-III) and 8.1% (Mat-II). The aforementioned also tended to more frequently return to the same student than teachers with a low share of complete-cycle implementation.

We also noted with regard to Tec-I a high number of exchanges (8C) which included successive cycles with a student also being accomplished by other student(s) interventions. This is another indicator of students' role in classroom FA.

Discussion

We discuss the results with respect to the research questions and context, but also in relation to the IFA and IBE literature in science education.

IFA implementation

The ESRU model used to analyse a classroom discussion in different STEM-inquiry lessons helps disclose differences in teachers' IFA practices.

All teachers in our study frequently elicited the students, recognised their responses and situated them with respect to the teaching goals (>60% of ESR cycle). However, situating students in relation to teaching goals does not imply that students' responses will be 'used' every time or immediately by teachers or students in a formative way to guide students' learning towards those goals. Nevertheless, it helps teachers establish where the students are in their learning process, maybe leading to them making better future teaching decisions or with stronger foundations (Black & Wiliam, 2009).

Moreover, our results show that the teachers major in the topic they teach and with long teaching experience, except for Mat-II, engage in different IFA practices. It is worth noting that many of them managed to make frequent use of effective IFA practices, in which students also participated. However, two teachers (PC-III and especially Mat-II) seemed more reluctant than the others to 'use' right approach to the collected information in order to guide the students' learning, hence they may have been lacking the strategies or supports needed for an on-the-fly FA (Cisterna & Gotwals, 2018; Harisson et al., this special issue).

It is noted that the literature indicates that only one out of three teachers implement a frequent (35%) effective ESRU complete cycle (the two other teachers <20%) while teaching physics topics (Ruiz-Primo & Furtak, 2007). Surprisingly, this teacher was an English literature major with a Master's degree in education, who was in her third year of teaching in grade 6, but her first year teaching science, compared to two other grade 7 teachers in the study, one holding a major in science and the other 14 years' teaching experience. The authors conclude that effective classroom IFA practices depend less on teachers' content knowledge and years of teaching experience than on teaching approaches and contextual factors.

Still, the teachers observed to frequently implement complete cycles tended more than the others to elicit students with more demanding cognitive and complex questions (e.g. 'formulate an explanation' and 'Compare/contrast another's idea') which goes beyond simply looking for the 'right answers'. Conversely, teachers with a low share of

Table 6. Examples of frequency of eliciting questions per category within a dimension and the use of strategies.

Eliciting		Mat-I	PC-I	Tec-I	Mat-II	PC-III
Epistemic	Formulate explanation /provide evidence and examples /relate evidence to explanation	41%	50%	27.4%	21%	29.3%
	Evaluate quality evidence / Compare-contrast observation-data-procedure/ Interpret information-data-patterns	2.1%	2.6%	4.8%	4.8%	4.9%
	Check comprehension / Apply known procedure	3,0%	2.6%	4.8%	17.7%	12.2%
	Make (Choose) Prediction / Make vote	14.7%	–	–	20.2%	14.6%
	Compare or contrast other's idea / promote debate	7.4%	15.8%	25.8%	5.6%	2.4%
Conceptual	Apply compare contrast concepts /Check comprehension	10.5%	15.8%	6.4%	1.6%	2.4%
Using						
Compares-Contrast students' responses /Promotes debating-discussion		13.7%	10.5%	6.4%	8.9%	4.9%
Promotes exploration idea /Promotes thinking (by asking Why/how?)		15.8%	31.6%	12.9%	4%	19.5%

complete-cycle implementation tended to focus more on checking students' 'simple' comprehension or applying known procedure than other teachers, ending in questions regarding the 'correctness' of the answers (e.g. see Mat-II/Table 2, 3T1: '... I need another argument because you may be wrong twice ... ').

Further, the 'Using' strategies (see Table 6) like 'Compare/contrast students' responses' and 'Promote debating/discussion', considered somewhat essential by the ESRU model for engaging students in assessment conversations, was more common among teachers with a high frequency of complete-cycle implementation-Mat-I (13.7%) and PC-I (10.5%)- and less by the teachers with the lowest ones-Mat-II (8.9%) and PC-III (4.9%).

However, the 'Using' strategies such as 'Promote thinking (asking Why/how?)' and 'Promote exploration of an idea' were applied differently by teachers and much more so by PC-I (31.6%) than Mat-II (4%).

Our results are consistent with literature conclusions on the characteristics of effective IFA practices attributed to teachers, namely:

- frequent interactions with students, complex and higher cognitive questions (Gotwals & Birmingham, 2016);
- engaging students in questioning, involving debating various responses and assessing students' progress in understanding the key lesson goals (Treagust et al., 2001) as opposed to teachers' convergent questions and searching for acceptable answers (Ateh, 2015); and
- using the information collected to make critical decisions on what to do next to guide students toward the learning goals (Cisterna & Gotwals, 2018; Ruiz-Primo & Furtak, 2007).

Hence, to better grasp the teachers' practices we situate our results in the context of the broader 'instructional environment' (Grangeat, 2015).

Our results are probably further strengthened by the fact that the teachers had the opportunity to participate in a three-year research project (except for Mat-II, a temporary beginner teacher), which included successive trials dedicated to implementing FA practices and IBE in the classroom. In addition, the teachers were engaged in communities of practice, accompanied and supported by tools for designing, adjusting and implementing FA key strategies (e.g. using precise self-assessment rubrics to make the learning goals

Table 7. Teacher frequency of IFA complete and incomplete cycles by inquiry dimensions.

	Conceptual dimension frequency				Epistemic dimension frequency						
	ES	ESR	ESRU cycle frequency	Total (%)	ES	ESR	ESRU	Total (%)	NI (%)	Total	Going back to the same student
Mat-I	4	2	4 1[3]M 1[9]C	10 (12.0%)	9	29 1[2] 6[5]	35 5[2] 1[3] 1C + 1[4] 1[11]	73 (87.9%)	12 (12.6%)	95	4[2] 2[3]
PC-I	1	3	4 1[3]M	7 (20.5%)	4	8 1[5]	15 1C + 2[4] 3[5]	27 (79.4%)	4 (10.5%)	38	1[2] 2[3]
Tec-I		2	1 2[7]M	3 (7.3%)	5	13	20 2C + 2[2] 1C[3] 2C[4] 2C[5] 1C[6]	38 (92.6%)	21 (33.8%)	62	1[2] 2[3]
Mat-II	1 1[2]	2	2 1[2]M	5 (5.6%)	10	58 1C[2] 2[5]C	19 2[2] 1[3] 1M + 2[4]	84 (94.3%)	32 (26.4%)	124	2[2] 1[3] 1[4]
PC-III		1	– 1C[8]M	1 (2.7%)	2	24	10 2[2] 1[3] 2C[5]	36 (97.2%)	4 (11.4%)	41	2[2]

Notes: Numbers in the second line in brackets [] represent the number of consecutive cycles (≥2 iterations) with the same student, numbers outside the brackets [] represent the frequency of these consecutive cycles. M represents cycles that include eliciting both conceptual and epistemic inquiry, C represents consecutive cycles that include to the same student interventions of other students'. In the last column, numbers inside the brackets [] represent the number of time the teachers go back to the same students, number outside brackets [] represent the frequency.

and criteria for success explicit; teachers providing feedback, through written FA and one-to-one IFA, that may move learners forward; and activating students as instructional resources for one another) in their classroom (Black & Wiliam, 2009; Cizek, 2010), which eventually prepared them for the IFA practices during the whole-classroom discussion. Moreover, the nature of the whole-classroom discussion may have given the possibility to make the students' thinking 'visible' through presentations and discussions and thereby offered the teachers (or students) greater opportunities for FA (Black & Wiliam, 2003).

The literature mentions that frequent FA practices may be associated with low socio-economic schools, and low FA practices with a high socio-economic context (Chen et al., 2012). Moreover, the discipline being taught may influence FA practices (e.g. mathematics teachers outperforming science ones') (Gotwals et al., 2015). Conversely, our study results in the context of low socio-economic schools show frequent and qualitative IFA with most teachers from different disciplines and that the science and technology teachers may have IFA which is comparable (or even surpasses that of) the mathematics teachers. Hence, other elements might also interfere with IFA (e.g. teachers' professional knowledge, teaching experience specifically regarding IFA practices, teacher's relationship with the community of practice, classroom context, etc.).

For example, PC-III is an experienced teacher, less enthusiastic in the preceding year's seminars for assessing lessons and making modifications, faced constraints with the topic taught (e.g. suggesting to change the teaching topic on the same day the researcher came to film the lesson and rushing to finish the course). In contrast, PC-I had the longest teaching experience in the group (>20 years), held a PHD in physics and was engaged in the research project (e.g. proposed tools for the lesson plan). PC-I also had no 'management' issues in her classroom, 'knew' the students' difficulties with regard to the topic under discussion and engaged them in conversations. Mat-I was motivated for the project (namely, her second project with the research team, and participated in the mathematics' lesson-plan preparation), knew the students' difficulties concerning the teaching goals and did not have an issue with managing her classroom.

Further, we adjusted the ESRU model to shed more light on the complexity of exchanges occurring during the IFA, in particular regarding the students' role in elaborating the IFA cycles. The results show that for two teachers (Tec-I and PC-II) the students played a significant role in the IFA practices. For example (see Table 3), the student peer-assessment that referred to a bridge design's fragility at a specific point (3Dri: 'it's falling') offered an opportunity for the presenting group to grasp the design's limitations, and the feedback given would probably allow them to improve their future design (GR1: 'yeah, later we just have to reinforce it here'). However, not all of the students' interventions in the ESRU phases are shown in Table 4, only the last phase of a cycle. These results still shed light on students' involvement in their own learning and in being an instructional resource for another (Black & Wiliam, 2009). In contrast, the way the students' interventions are perceived by their peers as a 'valuable' resource for learning (or not) could also influence the whole learning.

The literature indicates that students usually show a lack of confidence while evaluating the skills of their peers (Papinczak et al., 2007). Conversely, peers may be perceived to provide useful help when they are 'stuck' or need reassurance that their thinking was 'right' (Cowie, 2005). Moreover, feedback from one's peers, especially friends, is preferred to that of teachers. Indeed, students positively accept the (peer)assessment method when they think it contributes to their learning and when the assessment tasks are embedded in the regular course work, optimising peer-involvement (Tillema et al., 2011). These findings were also supported in our study, via integrating into lesson plans: teamwork, one-on-one IFA, a self-assessment grid and whole-classroom discussion, as supported by teachers and peers.

This manuscript does not explicitly consider whether students' learning will occur as an outcome of IFA practices. However, a detailed analysis of student learning might be included in further studies. Some literature results that addressed this particular issue using pre/post-tests seem to suggest learning does occur (Ruiz-Primo & Furtak, 2007), while in some cases the results were comparable to or exceeded those of students in comparative classrooms (Hickey et al., 2012). Still, it also seems that even the best-designed interventions do not always result in all students learning better (Black & Wiliam, 2009).

Intermediate level

Our results indicate that teachers with frequent complete cycle use tended to have longer exchanges with students and returned to the same student(s). Moreover, Tec-I's repeatedly eliciting the same student and allowing interference by others also led to longer exchanges.

These teaching strategies suggest that teachers took the 'necessary' time to situate the students regarding the learning goals and to move them forward in their learning. The literature indicates these teaching strategies may result in learning improvements (Ruiz-Primo & Furtak, 2007). For example, the why/when questions (see Table 1, 'So why does the temperature keep rising like that?'; 'When does it stop?') helped shed light on the students' reasoning and maybe deepened it ('8Margo: Water is going to be transformed into steam'). Yet, the student seems to be more answering the *when* question. It took the teacher several cycles and even going back later and again questioning the team to obtain the students' answers (S: 'Because of heat') to the *why* question, which seems to be at a higher level of complexity for students.

In contrast, long exchanges and returning to the same students at the expense of others may be against the improvement of 'all' students' learning. Studies show that the quantity of extended sequences bears no relationship with the quality of the discourse contained within them (Chen et al., 2012). Efficient IFA practices should allow teachers to hear from several students and focus on collective learning by synthesising ideas across students, instead of simply responding to a single student (Gotwals & Birmingham, 2016). In our research, teachers that made extensive use of complete cycles also relied on strategies like 'Compare/contrast ideas' and 'promote debate', which imply more collective teaching approaches and ultimately better learning.

Inquiry dimensions

The eliciting phase of the complete and incomplete cycles was mainly dominated by the epistemic inquiry dimension. These results suggest that teachers from different disciplines are more likely to deal with epistemic dimensions than conceptual ones. Further, the results show that teachers who frequently implemented complete cycles tended to use proportionally more conceptual inquiry dimensions than teachers who implemented them less often (PC-III and Mat-II).

Still, our results might be partly attributed to the nature of the session being analysed, focusing on argumentation, and because discursive practices are explicitly embedded in the epistemic dimension (e.g. 'Construct explanations'). Accordingly, in a further study, it would be important to verify whether the trends found in a whole-classroom discussion session are also noticeable in other sessions and other teacher–student(s) interactions (e.g. in small groups/one-on-one interaction).

It is also indicated in the literature that teachers implement epistemic dimensions much more than conceptual ones with regard to different physics lessons and that these two might overlap each other (Ruiz-Primo & Furtak, 2007). The literature also shows that teachers' inquiry enactments may vary especially in relation to each discipline (physics/biology/chemistry/earth-science) they teach, rather than to the classroom context (Breslyn & McGinnis, 2012). These results differ from ours, in which no specific disciplinary trend could be established between teachers from different disciplines and topics taught. Yet, our results are based more on general (across-discipline) inquiry dimensions and on analysing teachers' actual teaching enactments in the classroom via video recordings, differently from a study whose results relate more to specific-discipline(s) inquiries and are based on teachers' classroom portfolio declarative teaching.

Conclusion and implications

In this research, we analysed IFA practices and inquiries during a whole-classroom discussion session as implemented by STEM lower secondary teachers engaged in a research educational project. We managed to identify differences in IFA practices. Indeed, most teachers (except two) often implemented effective IFA practices, described by frequent complete ESRU cycle use (>33%), which was partially accomplished together with the students. Moreover, teachers tended to use the epistemic inquiry dimensions more than the conceptual ones'. Hence, teachers with a high frequency of ESRU cycles had a higher tendency to keep eliciting the same students at different times of the session while using longer consecutive cycles than those with a lower frequency of complete ESRU cycles.

These results imply that policies or research seeking to implement new educational approaches or teacher practices (e.g. IFA methods) in the classroom should concentrate on developing teacher professional knowledge by offering teachers adequate support for a suitable period of time. The literature states that IFA practices may be improved when (candidates or experienced) teachers are adequately supported, even for a relatively short period of time (less than one school year) (Gotwals et al., 2015; Gotwals & Birmingham, 2016; Hickey et al., 2012). As shown by Grangeat (2015), such support gives teachers the opportunity to access a rich environment that combines the theoretical background and teacher-specific needs (e.g. pedagogical content knowledge) in a comprehensive social context that allows the teacher to participate in a community of practice by conceiving, implementing, evaluating and adjusting in an iterative way specific teaching sequences that address the desired goals.

Note

1. Technology education in France is conceived so as to ensure continuity between lower/higher secondary and higher education and to prepare students to enter engineering colleges' and Masters of Science for engineers.

Acknowledgment

Moreover, the authors address their acknowledgments to Dr. Aikaterini Bougiatioti for her feedbacks on the manuscript.

Disclosure statement

No potential conflict of interest was reported by the author(s).

Funding

This work was supported by the 7th Framework Programme for Research and Technological Development (FP7), European union's Research and Innovation funding programme, project in education Assess Inquiry in Science, Technology and Mathematics Education (ASSIST-ME) [grant number 321428].

ORCID

Elie Rached http://orcid.org/0000-0002-1371-8965
Michel Grangeat http://orcid.org/0000-0002-1951-6547

References

Abd-El-Khalick, F., BouJaoude, S., Duschl, R., Lederman, N. G., Mamlok-Naaman, R., Hofstein, A., Niaz, M., Treagust, D., & Tuan, H. (2004). Inquiry in science education: International perspectives. *Science Education*, *88*(3), 397–419. https://doi.org/10.1002/sce.10118

Allal, L., & Mottier Lopez, L. (2005). Formative assessment of learning: A review of publications in French. In *Formative assessment – improving learning in secondary classrooms* (pp. 241–264). OECD Publication. https://archive-ouverte.unige.ch/unige:17429

Artigue, M., & Blomhøj, M. (2013). Conceptualizing inquiry-based education in mathematics. *ZDM Mathematics Education*, *45*(6), 797–810. https://doi.org/10.1007/s11858-013-0506-6

Ateh, C. (2015). Science teachers' Elicitation practices: Insights for formative assessment. *Educational Assessment*, *20*(2), 112–131. https://doi.org/10.1080/10627197.2015.1028619

Bernholt, S., Rönnebeck, S., Ropohl, M., Köller, O., & Parchmann, I. (2013). *Report on current state of the art in formative and summative assessment in IBE in STM – part I.* (Research Report No. 1. ISSN: 2246-2325). Department of Science Education, University of Copenhagen.

Black, P., & Wiliam, D. (2003). 'In praise of educational research': Formative assessment. *British Educational Research Journal*, *29*(5), 623–637. https://doi.org/10.1080/0141192032000133721

Black, P., & Wiliam, D. (2009). Developing the theory of formative assessment. *Educational Assessment. Evaluation and Accountability*, *21*(1), 5–31. https://doi.org/10.1007/s11092-008-9068-5

Breslyn, W., & McGinnis, J. R. (2012). A comparison of exemplary biology, chemistry, earth science, and physics teachers' conceptions and enactment of inquiry. *Science Education*, *96*(1), 48–77. https://doi.org/10.1002/sce.20469

Brown, J. (2012). The current status of STEM education research. *Journal of STEM Education*, *13* (5), 7–11. https://www.jstem.org/jstem/index.php/JSTEM/article/view/1652

Cazden, C., (2001). *Classroom discourse: The language of teaching and learning. Variations in lesson structure* (pp. 53-79). Portsmouth: Heinemann.

Chen, C.-H., Crockett, M. D., Namikawa, T., Zilimu, J., & Lee, S. H. (2012). Eighth grade mathematics teachers' formative assessment practices in SES-different classrooms: A Taiwan study. *International Journal of Science and Mathematics Education*, *10*(3), 553–579. https://doi.org/10.1007/s10763-011-9299-7

Cisterna, D., & Gotwals, A. W. (2018). Enactment of ongoing formative assessment: Challenges and opportunities for professional development and practice. *Journal of Science Teacher Education*, *29*(3), 200–222. https://doi.org/10.1080/1046560X.2018.1432227

Cizek, G. J. (2010). An Introduction to formative assessment: History, characteristics, and challenges. In H. L. Andrade & G. J. Cizek (Eds.), *Handbook of formative assessment* (pp. 3–17). Routledge.

Cobb, P., Confrey, J., DiSessa, A., Lehrer, R., & Schauble, L. (2003). Design experiments in educational research. *Educational Researcher*, *32*(1), 9–13. https://doi.org/10.3102/0013189X032001009

Cowie, B. (2005). Student commentary on classroom assessment in science: A sociocultural interpretation. *International Journal of Science Education*, *27*(2), 199–214. https://doi.org/10.1080/0950069042000276721

Driver, R., Newton, P., & Osborne, J. (2000). Establishing the norms of scientific argumentation in classrooms. *Science Education*, *84*(3), 287–312. https://doi.org/10.1002/(SICI)1098-237X(200005)84:3<287::AID-SCE1>3.0.CO;2-A

Duschl, R. (2008). Science education in three-part Harmony: Balancing conceptual, epistemic, and social learning goals. *Review of Research in Education*, *32*(1), 268–291. https://doi.org/10.3102/0091732X07309371

Duschl, R. A., & Gitomer, D. H. (1997). Strategies and challenges to changing the focus of assessment and instruction in science classrooms. *Educational Assessment*, *4*(1), 37–73. https://doi.org/10.1207/s15326977ea0401_2

Duschl, R. A., & Osborne, J. (2002). Supporting and promoting argumentation discourse in science education. *Studies in Science Education*, *38*(1), 39–72. https://doi.org/10.1080/03057260208560187

Gotwals, A. W., & Birmingham, G. (2016). Eliciting, Identifying, interpreting, and responding to students' ideas: Teacher candidates' Growth in formative assessment practices. *Research in Science Education*, *46*(3), 365–388. https://doi.org/10.1007/s11165-015-9461-2

Gotwals, A. W., Philhower, J., Cisterna, D., & Bennett, S. (2015). Using video to examine formative assessment practices as Measures of Expertise for mathematics and science teachers. *International Journal of Science and Mathematics Education*, *13*(2), 405–423. https://doi.org/10.1007/s10763-015-9623-8

Grangeat, M. (2015). Exploring the set of pedagogical knowledge, from pedagogy to content. In M. Grangeat (Ed.), *Understanding science teachers' professional knowledge growth* (pp. 117–133). Rotterdam: Sense Publishers. https://doi.org/10.1007/978-94-6300-313-1_7

Harlen, W., Artigue, M., Justin Dillon, J., & Léna, P. (2012, December). *Learning through inquiry background resources for implementing inquiry in science and mathematics at school*. Research Report. http://www.fibonacci-project.eu/

Hazelkorn, E., Ryan, C., Beernaert, Y., Constantinou, C., Deca, L., Grangeat, M., & Welzel-Breuer, M. (2015). *Science education for responsible citizenship* (Report No. EUR 26893). European Commission - Research and Innovation.

Hickey, D., Taasoobshirazi, G., & Cross, D. (2012). Assessment as learning: Enhancing discourse, understanding, and achievement in Innovative science curricula. *Journal of Research in Science Teaching*, *49*(10), 1240–1270. https://doi.org/10.1002/tea.21056

Lebeaume, J. (2011). *L'investigation pour l'enseignement des sciences: actualités des enjeux*. Dans Grangeat, M., (dir.) Les démarches d'investigation dans l'enseignement scientifique Pratiques de classe, travail collectif enseignant, acquisitions des élèves. Ecole Normale Supérieure de Lyon: Lyon, 19–34.

Mehan, H. (1979). What time is it, Denise? 'Asking known information questions in classroom discourse'. *Theory Into Practice*, *28*(4), 285–294. https://doi.org/10.1080/00405847909542846

Mercer, N., Dawes, L., Wegerif, R., & Sams, C. (2004). Reasoning as a scientist: Ways of helping children to use language to learn science. *British Educational Research Journal*, *30*(3), 359–377. https://doi.org/10.1080/01411920410001689689

Ministère de l'Education Nationale. (2008). Programme du Collège [Lower secondary programs]. *Bulletin officiel special*, n° 6 du 28 août 2008.

Mortimer, E., & Scott, P. (2003). *Meaning making in secondary science classroom*. Open University Press.

National Research Council. (2015). *Guide to implementing the next Generation science Standards*. The National Academies Press.

Osborne, J., Erduran, S., & Simon, S. (2004). Enhancing the quality of argumentation in school science. *Journal of Research in Science Teaching*, *41*(10), 994–1020. https://doi.org/10.1002/tea.20035

Papinczak, T., Young, L., & Groves, M. (2007). Peer assessment in problem-based learning: A qualitative study. *Advances in Health Sciences Education*, *12*(2), 169–186. https://doi.org/10.1007/s10459-005-5046-6

Rudolph, J. N. (2000). Reconsidering the `nature of science' as a curriculum component. *Journal of Curriculum Studies*, *32*(3), 403–419. https://doi.org/10.1080/002202700182628

Ruiz-Primo, M. A., & Furtak, M. E. (2007). Exploring teachers' informal formative assessment practices and students' understanding in the context of scientific inquiry. *Journal of Research in Science Teaching*, *44*(1), 57–84. https://doi.org/10.1002/tea.20163

Ruiz-Primo, M.-A., Furtak, E.-M., Ayala, C., Yin, Y., & Shavelson, R. (2010). Formative assessment, Motivation, and science learning. In H. Andrade & G. Cizek (Eds.), *Handbook of formative assessment* (pp. 139–158). Routledge.

Scott, P. H., Mortimer, E. F., & Aguiar, O. G. (2006). The Tension between Authoritative and dialogic discourse: A Fundamental Characteristic of Meaning making interactions in high school science lessons. *Science Education*, *90*(4), 605–631. https://doi.org/10.1002/sce.20131

Tiberghien, A., Malkoun, L., Buty, C., Souassy, N., & Mortimer, E. (2007). Analyse des savoirs en jeu en classe de physique à différentes échelles de temps. In G. Sensevy & A. Mercier (Eds.), *Agir ensemble: Eléments de théorisation de l'action conjointe du professeur et des élèves* (pp. 93–122). PUR.

Tillema, H., Leenknecht, M., & Segers, M. (2011). Assessing assessment quality: Criteria for quality assurance in design of (peer) assessment for learning – a review of research studies. *Studies in Educational Evaluation*, *37*(1), 25–34. https://doi.org/10.1016/j.stueduc.2011.03.004

Treagust, D., Jacobowitz, R., Gallagher, J., & Parker, J. (2001). Using assessment as a guide in teaching for understanding: A case study of a middle school science class learning about Sound. *Science Education*, *85*(2), 137–157. https://doi.org/10.1002/1098-237X(200103)85:2<137::AID-SCE30>3.0.CO;2-B

Wong, V., & Dillon, J. (2019). Crossing the boundaries: Collaborations between mathematics and science departments in English secondary (high) schools. *Research in Science & Technological Education*. https://doi.org/10.1080/02635143.2019.1636024

OPEN ACCESS

Making learning effective – quantity and quality of pre-service teachers' feedback

Mathias Ropohl and Silke Rönnebeck

ABSTRACT

Feedback is an essential formative assessment practice that has the potential to influence student learning and achievement positively. Providing effective feedback, however, is a challenging task for teachers. Especially beginning teachers struggle with the provision of information that supports students in developing and improving their competences. Learning to provide feedback thus is an important aspect of teacher education. The present exploratory study investigates pre-service chemistry teachers' practices regarding the judgement of students' level of achievement and the provision of feedback in the context of the control-of-variables strategy. A sample of $N = 40$ bachelor and master students at Kiel University judged students' written artefacts and provided feedback with the aim to support students in their progression towards the learning goals. The quality of the feedback was coded with respect to its correct judgement of the students' current levels of achievement and the support it provided regarding next steps in learning. The results show that only a few of the pre-service teachers could correctly judge students' current levels of achievement and provide feedback that is expected to be effective. Overall, the findings indicate a need for more research in this field and a need for the implementation of formative assessment practices as a topic in pre-service teacher education.

Introduction

With regard to the quality of instruction in science and other domains, one focus of educational research is the teacher and his practices in class (Geller, Neumann, & Fischer, 2014; Seidel & Shavelson, 2007; Vieluf, Kaplan, Klieme, & Bayer, 2012). Science teachers need professional knowledge in order to plan, conduct and analyse their teaching and adapt it to their students' needs (Baumert & Kunter, 2013; Gess-Newsome, 2015). The adaptation of teaching to students' individual learning requires teachers to realise formative assessment (Bell & Cowie, 2001; Black & Wiliam, 1998; Cizek, 2010). Teachers' practices that are related to formative assessment like interpreting students' ideas or providing feedback are relevant for successful student learning (Furtak et al., 2016; Gotwals & Birmingham, 2016; Millar, 2017). The proficient enactment of formative assessment practices by a teacher can lead to comparatively large gains in student achievement (Black & Wiliam, 1998; Hattie & Timperley, 2007; Kingston & Nash, 2011).

The present study focuses on two important aspects of formative assessment practices, on pre-service teachers' judgement accuracy and on the quantity and quality of their written feedback. Providing feedback is some kind of a bridge between teachers' intended learning goals and students' current levels of achievement in relation to those learning goals (Hattie & Timperley, 2007). Analysing pre-service teachers' judgements and the feedback they give based on this judgements, can provide insights into pre-service teachers' formative assessment practices.

Theoretical background

In this section, first an overview about (pre-service) teachers' formative assessment practices is given. A second part focuses specifically on feedback whereas the third part, eventually, describes the control-of-variables strategy (CVS) thus setting the scene for the science context in which this study took place.

(Pre-Service) teachers' formative assessment practices

Definitions of formative assessment emphasise the importance of continuous assessments for student learning: 'Formative assessment, also referred to as assessment for learning, is a process in which teachers and students recognise and respond to student learning, during that learning' (Cowie, 2012, p. 679). Gotwals and Birmingham (2016) identified four practices that together describe a well enactment of formative assessment on the teachers' side: Teachers (1) 'provide opportunities for students to demonstrate their understandings', (2) 'identify students' ideas', (3) 'interpret students' responses', and (4) 'respond to students' ideas with feedback and/or instructional decisions that will help move from or build on current understandings' (p. 366; see also Furtak et al., 2016; Haug & Ødegaard, 2015). Research in the German speaking countries has tended to focus more on a specific aspect of these practices called teachers' judgement accuracy (Anders, Kunter, Brunner, Krauss, & Baumert, 2010; Schrader, 2013). Judgement in this case means that the teacher has to interpret accurately what a piece of student work (i.e. what was said, written or created) implies with respect to the achievement of specific learning goals or the progress towards these goals (Dolin, Black, Harlen, & Tiberghien, 2018). Accurately involves that the teacher uses criteria to judge the student work and that it is possible to quantify somehow the judgement, i.e. based on rubrics (Panadero & Jonson, 2013). Thus, judgement accuracy is an integral part of teachers' formative assessment practices since the identification and interpretation of student ideas and responses also includes a judgement. The practices, however, stress the relevance to enact a particular response much more than judgment accuracy does as they reflect the process-like character of formative assessment and as they integrate feedback into this process (Cowie, 2012; Furtak et al., 2016).

Consequently, formative assessment practices are an essential part of teachers' professional competence. Models of teachers' professional competence reflect this importance by defining facets that encompass formative assessment practices (Baumert & Kunter, 2013;

Gess-Newsome, 2015; Park & Oliver, 2008). Some models regard assessment knowledge as a generic aspect of teachers' professional knowledge base (Gess-Newsome, 2015), others as a content specific aspect that is part of their pedagogical content knowledge (PCK; Baumert & Kunter, 2013; Park & Oliver, 2008). In both cases, judgement accuracy and feedback are measureable products of teachers' formative assessment practices that help to approximate their professional competence.

In general, only a few empirical findings exist that focus on teachers' formative assessment practices in the science subjects. Two studies from mathematics and science education show that the judgement of students' levels of achievement regarding subject-specific concepts is comparably easy for in-service teachers (Heritage, Kim, Vendlinski, & Herman, 2009; Herman, Osmundson, & Silver, 2010). This is also true for pre-service teachers from the two domains who had to analyse students' responses to probes referring to two chemical phenomena (Talanquer, Bolger, & Tomanek, 2015). In contrast, in-service and pre-service teachers struggle in the same context with the interpretation of students' ideas without any support (Furtak et al., 2016; Nussbaum, 1981; Talanquer et al., 2015). In addition, teachers often tend to make general remarks when giving feedback that do not support next steps in student learning.

The reported studies have two things in common. First, they focus mainly on formative assessment practices related to students' conceptual knowledge. Consequently, hardly no empirical findings exist that focus on pre-service teachers' formative assessment practices in the field of scientific inquiry. Compared to the assessment of students' conceptual knowledge, inquiry competences require different forms of interpreting students' ideas. Second, feedback is often not in the focus of pre-service teachers' practices. Both limitations are addressed in the two following sections by summarising the research regarding the quantity and quality of feedback in general (and not as an integral part of formative assessment practices) as well as the research regarding the control-of-variables strategy (CVS) as an important aspect of scientific inquiry.

The quantity and quality of feedback

Providing effective feedback is a challenging task for pre-service teachers because it requires different types of knowledge like, e.g. content knowledge (CK), curriculum knowledge and assessment knowledge (Baumert & Kunter, 2013; Park & Oliver, 2008). Pre-service teachers must learn how effective feedback looks like with respect to its layout (e.g. form and timing) and which components of feedback are the decisive ones (e.g. focusing on next steps in learning). In several empirical studies and meta-analyses, the conditions and characteristics of feedback that positively affect learning were identified (Hattie & Timperley, 2007; Kluger & DeNisi, 1996) and transferred into guidelines for teachers how to give feedback (Brookhart, 2008).

Effective feedback should support students in closing the gap between their current levels of achievement and predefined learning goals (Hattie & Timperley, 2007). Thus, it is not only criterion-referenced but also linked to an individual reference norm (Dolin et al., 2018) and should provide students with answers to three major questions: 'Where am I going? (What are the goals?), How am I going? (What progress is being made toward the goal?), and Where to next? (What activities need to be undertaken to make better progress?)' (Hattie & Timperley, 2007, p. 86). Brookhart (2008) summarises characteristics of effective feedback related to both, its format and its content. Effective feedback should be descriptive rather than judgemental and focus on both, the outcome of the student work but also the processes leading to that outcome. Important aspects influencing its effectiveness are the timing and amount of the feedback as well as its clarity and specificity. Feedback should be given while the students are still aware of the learning goals and still have the time and opportunity to act on it. The amount and the content should be tailored to students' developmental levels; it might be necessary for the teacher to prioritise specific learning goals and focus on these in his or her feedback. Eventually, providing effective feedback resembles a tightrope walk. Teachers have to find a balance so that their comments are specific enough for students to know what to do next without giving away the right answer and doing the work for them. Whereas these criteria apply to all types of feedback, written feedback in particular requires clarity and specificity. Moreover, the tone and word-choice are of crucial importance. They should communicate respect for the students as active learners and 'inspire thought, curiosity, or wondering' (Brookhart, 2008, p. 34).

In general, there is only limited research on pre-service as well as in-service teachers' ability to give effective feedback and how that feedback addresses students' needs. However, the few existing findings give first insights into the strengths and weaknesses of pre-service teachers' feedback. One of the biggest challenges for both, pre-service and in-service teachers is the planning of next steps in teaching and learning based on diagnostic information (Buck, Trauth-Nare, & Kaftan, 2010; Heritage et al., 2009). In their analysis of in-service teachers' written feedback in students' science notebooks, Ruiz-Primo and Li (2013), e.g. found that teachers' feedback comments provided mostly either evaluative information about the quantity of work or the general level of understanding or descriptive information about how and why the work is right or wrong. Only four percent of the comments were prescriptive, i.e. provided information for students what to do next. Moreover, the teachers in general used comments much less frequently than grades in their feedback. Buck et al. (2010) report that pre-service teachers 'proved unable in most cases to provide students with substantive feedback that would have engaged students in active participation in their own learning' (p. 418).

The control-of-variables strategy

When thinking about feedback, it is important to focus on major subject specific learning goals and on empirical findings that report students' learning difficulties regarding these goals (cf. Hattie & Timperley, 2007).

In the past, research on feedback has mostly focused on scientific concepts and students' knowledge regarding these concepts. In contrast, competences that are related to the scientific inquiry have only marginally been considered although the respective competences are major subject specific learning goals in the science subjects. Scientific inquiry is a construct that requires a set of related competences like *planning and carrying out investigations* (e.g. NGSS Lead States, 2013). A fundamental principle of these investigations is to isolate the causal effects of variables by contrasting conditions that differ only with respect to the variable for which the causal status is assumed. The critical competence required for planning a controlled investigation is known as *control-of-variables strategy* (CVS). The application of this strategy in investigations can be characterised by five criteria: (1) defining and measuring the dependent variable, (2) identifying and varying the independent variable, (3) controlling possible confounding variables, (4) specifying the length of measurements, and (5) repeating measurements (Arnold, Kremer, & Mayer, 2014). Many current science curricula and standards address the CVS explicitly or implicitly (KMK, 2005; NGSS Lead States, 2013). It is regarded as a prerequisite for successful science learning in general and for successful scientific inquiry learning specifically as it enables students to conduct their own controlled investigations and to derive arguments from the results (Schwichow, Zimmerman, Croker, & Härtig, 2016). Against this background, the CVS can be described as a major subject specific learning goal.

Moreover, science education research has investigated students' competences and especially their difficulties when doing inquiry in the past (Rönnebeck, Bernholt, & Ropohl, 2016). A competence that has been extensively focused on is the planning of an investigation that tests given or self-generated hypotheses. Testing hypotheses in a valid way requires the application of the CVS (Chen & Klahr, 1999). However, students from different grades have difficulties when they have to apply the CVS (e.g. Arnold et al., 2014; Siler & Klahr, 2012). Typical errors are that students

(1) basically do not recognise the need for an experimental control,
(2) do not differentiate between variables to be tested and those that need to be controlled,
(3) do not take all possible influencing variables into account,
(4) vary the characteristics of the independent variable unsystematically,
(5) do not keep confounding variables constant, and
(6) do not verify the measurement by a repeated measurement.

Even specific hands-on or paper-pencil trainings on applying the CVS do not lead to its deeper conceptual understanding (Schwichow et al., 2016). This makes it necessary to analyse further possible factors that could support the development of the CVS (Schwichow, Croker, Zimmerman, Höffler, & Härtig, 2016). Results from a meta-analysis imply that feedback with a focus on the CVS could be an important factor that substantially supports students' development of the CVS (Ross, 1988).

Research aims and research questions

From a research perspective, the CVS is a good example to investigate pre-service teachers' formative assessment practices and especially the quantity and quality of their feedback in the context of scientific inquiry since teachers need to be aware of students' typical errors regarding the CVS, need to be able to diagnose their students' competences in connection with it, and need to give effective feedback. By focusing on these aspects, the exploratory study presented on the subsequent pages addresses pre-service teachers' formative assessment practices in a field that has not been sufficiently addressed in the past but that is fundamental for learning science.

The general aim is to generate empirical research findings that describe the two important practices of judging student work and providing effective feedback in connection with the CVS and its defining criteria. More specifically, the research intends to answer the following research questions. The first question is related to pre-service teachers' ability to judge students' application of the CVS:

- (F.1) To what extend are pre-service teachers able to judge students' levels of achievement with respect to the CVS and its defining criteria?

The second research question focuses on the quantity and quality of pre-service teachers' feedback regarding the CVS:

- (F.2) To what extent are pre-service teachers able to formulate feedback that has the potential to be effective for student learning with respect to the CVS and its defining criteria?

Method

Participants

In order to answer the research questions, a study was conducted that used a convenience sample of students from Kiel University. In total, $N = 40$ students formed the sample that was drawn in the winter term 2015/2016. The $n = 21$ bachelor and $n = 19$ master students were studying to become chemistry teachers at upper secondary schools in Germany. The mean number of terms studied was $M = 5.67$ ($SD = 1.93$) for the bachelor students and $M = 1.74$ ($SD = 1.82$) for the master students. For comparison, the bachelor programme takes six terms and the following master programme four terms (Neumann, Härtig, Harms, & Parchmann, 2017).

The structure of the teacher education programmes at Kiel University is representative for the structure of such programmes at other German universities (Bauer, Diercks, Rösler, Möller, & Prenzel, 2012). It involves that bachelor students predominantly attend courses that focus on the development of pre-service teachers' CK whereas master students mostly attend courses that focus on the development of their PCK (Neumann et al., 2017). Consequently, the master students of this sample had nearly finished all PCK courses of the entire programme and the bachelor students only about half of them. Since aspects of formative assessment practices are solely part of the PCK courses, master students could be expected to have a higher PCK regarding formative assessment practices than bachelor students. The bachelor students participated voluntarily in the study within a regular seminar. Each master student attended an extracurricular seminar at the university and received 50 € for his or her voluntary participation in this study.

Research design

The study of about 120 minutes consisted of three parts. In the first part, the test persons filled in a questionnaire asking for background variables (e.g. their subjects and their number of terms studied). In the second part, the CVS and general characteristics of effective written feedback were briefly thematised in a 20-minute research based presentation summarising the state of the art. The presentation explained the CVS and reported typical mistakes students make when they apply the strategy. Besides, the presentation focused on effective feedback. Based on the presentation and additional material, the pre-service teachers then, in the third part of the study, had to judge three authentic student artefacts with regard to the CVS and give feedback that shows characteristics of effective feedback.

The three artefacts – (1) UTJ18, (2) KIE21, and (3) MAF10 – were students' written plans for an experimental investigation. The artefacts were written by students from 8th grade within an intervention study. Students were asked to plan an investigation to test the hypothesis *The black layer* [that is formed when you heat a piece of copper foil in the flame of a Bunsen burner] *consists of soot that comes from the flame of the Bunsen Burner*, a hypothesis typically given by students in the context of the question *How is the black layer formed that appears if you heat a piece of copper foil in the flame of a Bunsen burner?* The question is related to the so-called copper problem (Hadfield, 1995) which aims at making students' misconceptions regarding chemical reactions visible.

Data collection

The pre-service teachers used a rubric for their judgements that encompasses ten characteristics of an investigation plan (Table 1). These characteristics are relevant with respect to the CVS and refer to the five criteria mentioned above (Arnold et al., 2014). For each of these characteristics, the test persons judged the degree to which the three written plans met them by marking the appropriate checkbox on the rubric (wrong/not answered, partially achieved, achieved). As some kind of a reminder, the test persons could use a separate sheet of paper that gave a definition of each characteristic.

Based on the described judgements, each pre-service teacher wrote a feedback text for each artefact. For that purpose, the test persons used a second sheet of paper that could possibly be given to students. It included a description of the desired learning goals,

Table 1. Description of ten characteristics of a written investigation plan.

	Judgement		
Characteristics	Wrong/not answered (code: 0)	Partially achieved (code: 1)	Achieved (code: 2)
Verification of the hypotheses	Student poses a hypotheses that cannot be proved.	Student poses a hypotheses that can be partly proved.	Student poses a hypotheses that can be fully proved.
Dependent variable	Student doesn't name dependent variable.	–	Student names dependent variable.
Measurement of dependent variable	Student does not observe/ measure the dependent variable.	Student observes dependent variable.	Student measures dependent variable.
Independent variables	Student doesn't name independent variable.	–	Student names independent variable.
Size/volume disclosure for the independent variable	Student does not consider the size/volume.	Student considers size/ volume indirectly.	Student considers size/ volume directly.
Controlled variation of independent variable	Student does not vary the independent variable.	Student varies the independent variable, but uncontrolled.	Student varies the independent variable controlled.
Consideration of confounding variables	Student does not consider confounding variables.	Student considers one or two confounding variables.	Student considers more than two confounding variables.
Repetition of experiment	Student does not describe a repetition.	Student describes a repetition with same OR other objects.	Student describes a repetition with same AND other objects.
Used materials	Student does not list materials.	Student lists some materials.	Students lists all materials.
Consideration of safety precautions	Student considers no safety precautions.	Student considers one safety precaution.	Student considers two or more safety precautions.

there was space for the judgement of the current learning level, and there was space for individual feedback. In their feedback, the participants were instructed to exemplify aspects that should be taken into account when planning a new investigation under controlled conditions.

In total, each of the $N = 40$ test persons judged the three different artefacts and gave written feedback based on that judgement. This resulted in $N = 119$ judgements and feedback texts (40 for each artefact with 1 missing for KIE21; 62 judgements by bachelor students, 57 judgements by master students, and 1 missing judgement by 1 bachelor student).

Data coding and data analysis for research question one

In order to answer the first research question, pre-service teachers' judgements of students' artefacts were scored and then compared with judgements given by experts. The scoring was based on the characteristics for a written investigation plan (Table 1). The scores for each characteristic (given by the marked checkboxes, i.e. wrong/not answered = score 0, partially achieved = score 1, achieved = score 2) were added up resulting in a total judgement score. In addition, the result of the scoring of each written plan was transferred to five distinct levels that describe different qualities of the application of the CVS (Table 2). Since the scoring as described here consisted of a simple addition of scores and required no interpretation of data, no interrater reliability needed to be considered. After that, the judgement accuracy was calculated by comparing pre-service teachers' judgement scores with the experts' judgement scores (see the next paragraph for a description of the expert coding) using Cohen's Kappa. In order to disclose the characteristics that were most difficult for the pre-service teachers to judge, the average absolute deviation between the experts' judgement and the pre-service teachers' judgements was analysed for each characteristic and for each artefact separately.

Based on the rubric presented in Table 1, two experts coded each artefact in order to generate a reference norm for the analysis of pre-service teachers' judgements. The two experts were university staff members working for the project and having a background in chemistry education. First, each expert coded the three artefacts independently. Next,

Table 2. Description of five levels that reflect the quality of the application of the CVS.

Level of quality	Description of level of quality	Characteristics that are relevant on that level	Students' achievement with respect to the characteristics[a]
5	The designed experiment varies the independent variable under strongly controlled conditions.	Controlled variation Consideration of confounding variables	2 2
4	The designed experiment varies the independent variable under semi-controlled conditions.	Controlled variation Consideration of confounding variables	2 1
3	The designed experiment varies the independent variable systematically.	Controlled variation Consideration of confounding variables	2 0
2	The designed experiment varies the independent variable, but unsystematically and uncontrolled.	Controlled variation Consideration of confounding variables	1 0
1	The designed experiment does not vary the independent variable.	Controlled variation Consideration of confounding variables	0 0

Note. [a]0 = wrong/not answered, 1 = partially achieved, 2 = achieved.

the two experts brought about 100% consensus by discussing their coding. A perfect experimental plan would get a maximum judgement score of 20 points. According to the expert coding, the first artefact (UTJ18) received a judgement score of 6 points, the second (KIE21) of 11, and the third (MAF10) of 13. According to the different levels that reflect the quality of the application of the CVS (Table 2), the first artefact was on level 1, the second on level 3 and the third also on level 3. In total, there was thus one poor (UTJ18), one average (KIE21) and one slightly better than average example (MAF10). The poor artefact is a description of an experiment under not systematically varied conditions where the student does not differentiate between variables to be tested and those that need to be controlled. The average artefact picks up the unusual idea of comparing the reaction behaviour of three different metals. Based on that idea, the student plans an appropriate experiment under controlled conditions. The best artefact reports a perfect idea but some aspects are missing: the influence of confounding variables, the definition of a specific volume for the independent variable, and a repeated measurement.

Data coding and data analysis for research question two

In order to answer the second research question, pre-service teachers' feedback texts (N = 119) were analysed. The texts were coded based on a coding manual using four codes (Figure 1). Two student workers who studied to become chemistry teachers in the master programme at Kiel University were trained before the coding. This training encompassed an introduction to the rubric and a test run. Afterwards, they did the coding independently. For this analysis, the feedback texts were divided into comments. A section of the text was counted as one comment if the pre-service teacher either referred to *one (or more) of the characteristics* of a written investigation plan in *one sentence* or if he or she referred to *one characteristic* (or general aspect of the student work) in *more than*

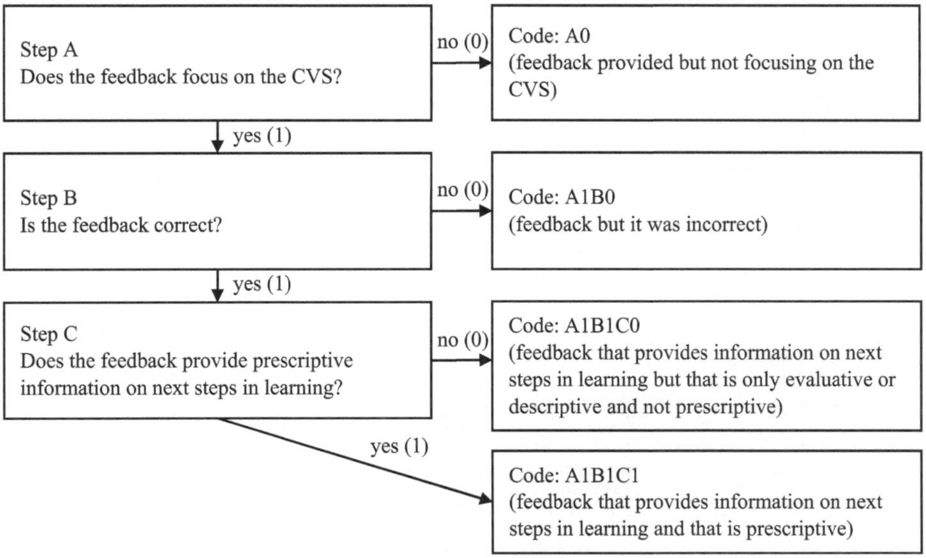

Figure 1. Flow sheet in order to code pre-service teachers' feedback comments.

one sentence ($N = 269$ comments). In order to calculate the accuracy of their coding, ten percent of the comments were randomly drawn from the basic population of comments (27 out of 269 comments) and then coded by both student workers. By chance, these 27 comments were only from the two most common categories A1B0 and A1B1C0 (Figure 1). Next, the interrater reliability was calculated using Cohen's Kappa. It attained a value of $\kappa = 0.74$ that indicates a substantial agreement between the two coders (Landis & Koch, 1977). Only for three of the 27 comments, the two coders did not agree.

Results

In order to answer the research questions, the collected data were analysed in two steps. First, pre-service teachers' judgement accuracy was analysed by comparing pre-service teachers' with experts' judgements. Next, the quantity and quality of pre-service teachers' feedback regarding the next steps in learning was examined.

Pre-service teachers' judgement accuracy

Pre-service teachers' judgement accuracy is displayed in Figures 2 and 3. The bar charts show the accuracy for the judgements by bachelor and master students separately categorised by the strength of the agreement. In total, there are four categories representing (1) a poor, (2) a moderate, (3) a substantial or (4) an almost perfect agreement (Landis & Koch, 1977).

Most judgements of the best artefact and nearly all judgements of the average artefact by the bachelor students have a poor agreement with the experts' judgement (Figure 2).

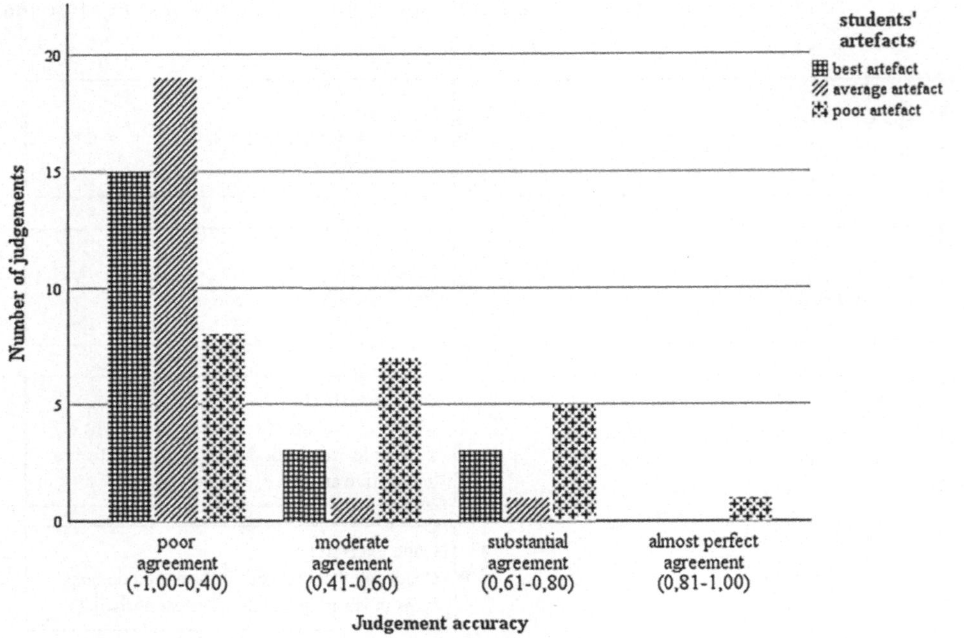

Figure 2. Categorised judgement accuracy for 62 judgements by bachelor students.

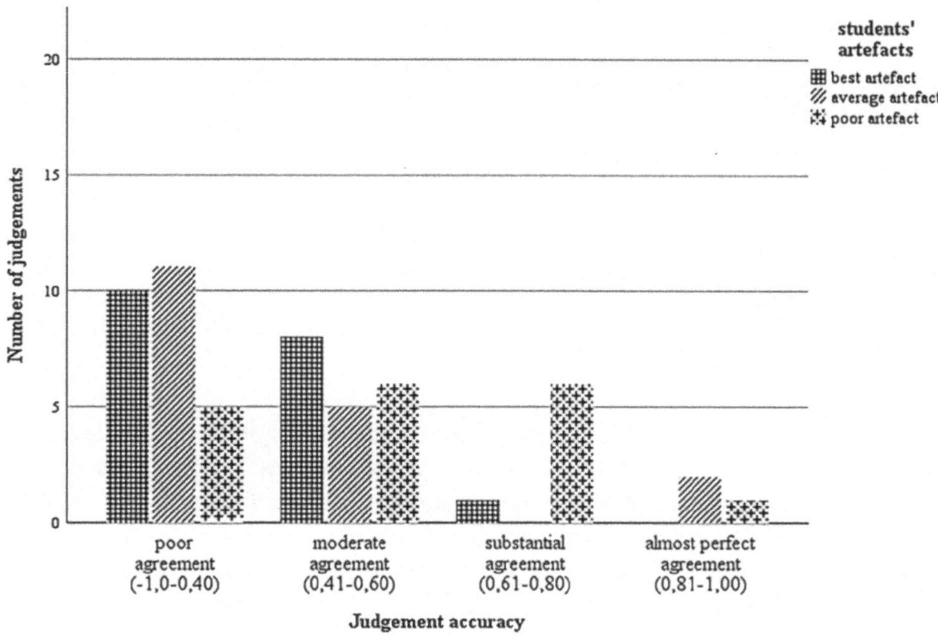

Figure 3. Categorised judgement accuracy for 57 judgements by master students.

Only for the poor artefact, the majority of the judgements has a moderate or substantial agreement. The situation is slightly different for the judgements of the master students (Figure 3). Compared to the bachelor students, there are more judgements of the best and average artefact that have a moderate agreement and less that have a poor agreement with the experts' judgement. Very few pre-service teachers, however, reach an almost perfect agreement.

Based on the codes given by the pre-service teachers and the experts, respectively, the three artefacts were in addition assigned to the five levels (Table 2) reflecting the quality of the application of the CVS. Table 3 shows the agreement between the experts' assignment (depicted in bold) and the distribution of assignments resulting from the judgements of the pre-service teachers. According to the expert judgement, the best artefact was coded

Table 3. Matching between the experts' and pre-service teachers' judgements of the levels that reflect the quality of the application of the CVS.

Best artefact (MAF10)		Average artefact (KIE21)		Poor artefact (UTJ18)	
Level[a]	Number of judgements	Level[a]	Number of judgements	Level[a]	Number of judgements
5	2	5	3	5	0
4	9	4	11	4	0
3	19	**3**	4	3	0
2	5	2	9	2	5
1	3	1	10	**1**	33
Missing	2	Missing	2	Missing	2
Total number of judgements	40	Total number of judgements	39	Total number of judgements	40

Note: [a]cf. Table 2. in bold = level assigned to artefact by expert rating.

Table 4. Average absolute deviation between the experts and the pre-service teachers' judgements.

Judgement criteria	Best artefact (MAF10) [*D*]	Average artefact (KIE21) [*D*]	Poor artefact (UTJ18) [*D*]
Verification of the hypotheses	0.20	**1.00**	**0.56**
Dependent variable	**0.45**	**0.80**	0.13
Measurement of dependent variable	0.39	**0.46**	**0.46**
Independent variables	**0.73**	**0.95**	0.08
Size/volume disclosure for the independent variable	**0.68**	**0.70**	0.21
Controlled variation of independent variable	0.28	**0.79**	0.13
Consideration of confounding variables	0.36	**0.95**	0.08
Repetition of experiment	**1.33**	**0.45**	**1.03**
Used materials	0.33	0.15	0.28
Consideration of safety precautions	0.33	0.30	**0.58**

Note: In bold = average absolute deviation $\geq$ 0.45.

on level three, which means that the experiment included a systematic variation of the independent variable. Nearly half of the pre-service teachers matched this result. A quarter of the pre-service teachers did not recognise this systematic variation or a variation at all (levels 2 and 1) whereas another quarter identified a consideration of confounding variables that was not there (levels 4 and 5). The average artefact was also coded by experts as being on level three.[1] Only approximately ten percent of the pre-service teachers matched this expert judgement. Half of the pre-service teachers coded the artefact on a lower level, and more than one third on a higher level. Obviously, the best and especially the average artefact were judged quite differently by different pre-service teachers. A higher agreement can be observed for the poor artefact. Here the expert judgement is matched by more than 80 percent of the pre-service teachers. Only 12.5 percent of the pre-service teachers coded the artefact on the next higher level.

Because of the obvious differences, a deeper look on the data was realised by looking at the deviation between the experts' judgement and the pre-service teachers' judgements (Table 4). Again, the analysis revealed that there is no general trend reflecting pre-service teachers' judgement accuracy. However, it becomes obvious that the average artefact is somehow challenging for the test persons as for eight out of ten characteristics the average absolute deviation is higher than 0.45. Regarding the other two artefacts, the average absolute deviation is higher than 0.45 for only four characteristics. In view of the characteristics, three issues seem to be challenging: (1) the definition of the dependent and independent variable, (2) the manipulation or the measurement of the variables, and (3) the repetition of the experiment. For the respective characteristics, the deviation is higher than 0.45 for at least two of the three artefacts.

Quantity and quality of pre-service teachers' feedback

In this section, all written comments given by the pre-service teachers are analysed qualitatively and quantitatively (Table 5). In total, there are $N = 269$ comments by $N = 40$ pre-service teachers. 111 comments refer to the poor artefact (UTJ18), 70 to the average artefact (KIE21) and 88 to the best artefact (MAF10). These comments were allocated to four categories describing their quality using the flow sheet in Figure 1. Based on the coding, about 10 percent of the comments inform the students about the next steps in learning (Table 5). These prescriptive comments are the comments with the code A1B1C1 (e.g. *In the future, please consider confounding variables so that you can keep the conditions*

Table 5. Distribution of feedback comments with respect to the quality of the student artefact and the quality of the feedback comments.

	Number of comments			
Nature of feedback comments[a]	Best artefact (MAF10)	Average artefact (KIE21)	Poor artefact (UTJ18)	Total
all comments	88	70	111	269 (100.0 %)
no focus on CVS (code A0)	10	13	17	40 (14.9 %)
incorrect (code A1B0)	32	18	14	64 (23.8 %)
descriptive or evaluative (code A1B1C0)	41	31	64	136 (50.6 %)
prescriptive (code A1B1C1)	7	10	12	29 (10.8 %)

Note: [a]cf. Figure 1.

of your experiment constant).[2] The majority of the comments is incorrect (e.g. *Identify the dependent and independent variable* although the student identified them correctly) or descriptive (e.g. *Identify the variables*) and evaluative (e.g. *You have not yet understood the definition of the variables*; *Good test*). 15 percent of the comments do not focus on the CVS (e.g. *Please, read the task more carefully*; *Sentence structure*). The prescriptive comments are distributed unevenly across the three artefacts: the poor artefact got the most and the best the least comments. Regarding the rather descriptive and evaluative feedback, most of the comments referred to the poor artefact. The average and the best artefact received only about half of the comments.

The comments are distributed differently across the different characteristics of a written investigation plan. Comments that are incorrect mainly refer to the definition of the dependent ($n = 15$; 23%[3]) and the independent variables ($n = 20$; 31%) or they focus on the verification of the hypotheses ($n = 16$; 25%). Feedback comments that are evaluative and descriptive also focus on the definition of the dependent ($n = 38$; 28%) and the independent variables ($n = 38$; 28%). In addition, a substantial number of comments refers to the definition of the confounding variables ($n = 42$; 31%). Other characteristics that are commented most are the verification of the hypotheses ($n = 20$; 15%), the controlled variation of the independent variable ($n = 27$; 20%) and the repetition of the experiment ($n = 28$; 21%). The analysed patterns resurface when looking at the results for the comments that are prescriptive. Again, the majority of the comments focuses on the definition of the dependent ($n = 13$; 45%), independent ($n = 13$; 45%) and the confounding variables ($n = 14$; 48%). Only very few comments referred to the other characteristics of a written investigation plan.

Looking at the feedback comments from a qualitative perspective, especially the incorrect comments (code A1B0) are interesting because they might provide insights into problems that pre-service teachers face. The analysis reveals different types of problems. Conceivably, most problems seem to be related to the pre-service teachers' own missing competence to apply the CVS. They have especially difficulties with the identification of the dependent ($n = 15$; 23%) and independent variable ($n = 20$; 31%; e.g. *Next time think carefully how you can describe your independent variable and dependent variable better and more clearly* although the description was very clear) and the manipulation or the measurement of the variables ($n = 6$; 9%; $n = 9$; 14%; e.g. *Is there more than one independent variable?*) which supports to some extend the findings presented in Table 5. Moreover, however, one can identify two more problems from the feedback comments. First, some of the pre-service teachers themselves seem to lack the necessary CK since they believe that the black layer is soot (e.g. *How can you proof that the soot really comes from the flame?*). A second problem is related more generally to the students' own inquiry competence. They have problems understanding the hypothesis and the aim of the experiments ($n = 16$; 25%; e.g. *Think again carefully if you can really find out with your experiment what the black layer is* – which was not the aim) or they do not realise that the unusual approach in the average artefact is a way to test the hypothesis (e.g. *Your design and your experiment are well structured but you do not test your hypothesis*).

On average, bachelor and master students give feedback to a different extent (Figure 4). The average number of comments per bachelor student is $M = 5.9$, whereas master students give $M = 7.3$ comments. The average number of comments is distributed differently across the four categories that reflect the quality of the feedback. Compared to bachelor students, master students give more comments that focus on the CVS, especially comments that reflect descriptive and evaluative or prescriptive feedback. Bachelor students, on the contrary, tend to give feedback comments that do not focus on the CVS.

Discussion of findings and conclusions

The aim of this exploratory study was to investigate pre-service teachers' judgement accuracy and their ability to provide feedback to students that is expected to be effective for learning in the context of the CVS in science. A sample of $N = 40$ bachelor and master students in chemistry education at Kiel University was asked to judge and provide feedback to three authentic written investigation plans of different quality.

With respect to the first research question, i.e. pre-service teachers' competence to accurately judge students' levels of achievement, the results show that their judgement accuracy is generally low. Only a few pre-service teachers reach a more than substantial

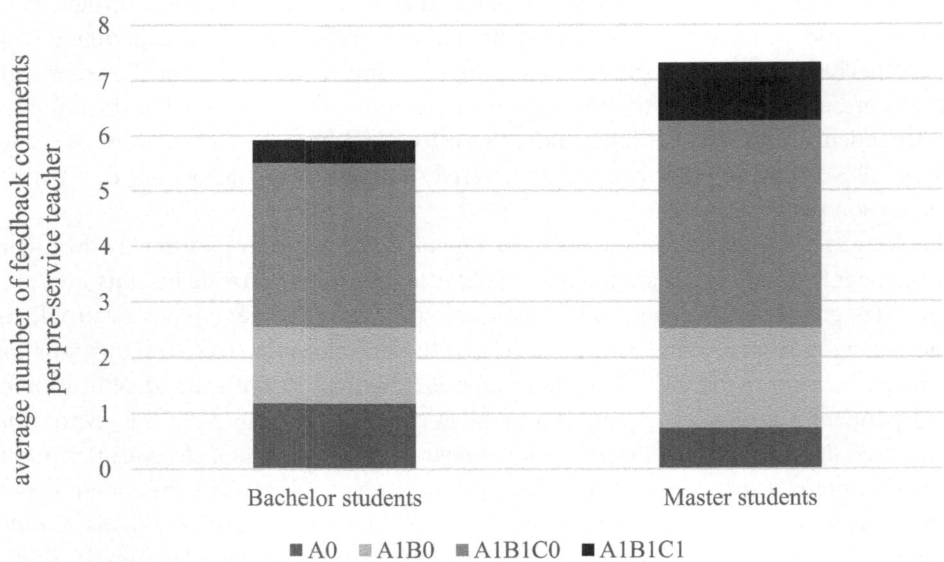

Figure 4. Average number of feedback comments per pre-service teacher.

agreement with an expert coding. The accuracy thereby depends on both, the study programme, bachelor or master, and the quality of the student work. Master students tend to reach a higher accuracy than bachelor students do. This finding reflects the structure of teacher preparation programmes at universities (Neumann et al., 2017). Master students have attended more courses on PCK than bachelor students have and thus should have better developed formative assessment practices. For both groups of students, it seems to be easier to judge a work of poor quality than of average or better than average quality which could be related to their own level of CK (see below). These findings help to not only differentiate previous research results that stressed the importance of supporting teachers' formative assessment practices to judge students' work (Heritage et al., 2009; Talanquer et al., 2015) but expand them from the field of conceptual knowledge to that of scientific inquiry.

Concerning the second research question, i.e. if pre-service teachers are able to formulate feedback regarding next steps in learning, the findings of this study show that only about one tenth of the feedback comments are prescriptive in nature. Half of the comments are descriptive or evaluative, 15 percent do not focus on the CVS and almost one quarter is incorrect. The domination of descriptive and evaluative as well as the small amount of prescriptive comments are well in-line with earlier research that focused either on in-service teachers (Heritage et al., 2009; Ruiz-Primo & Li, 2013) or pre-service teachers (Buck et al., 2010) in content-based contexts. It is also in line with findings indicating that science teachers do not put an emphasis on feedback in general (Schiepe-Tiska et al., 2016). In a similar vein, the high amount of incorrect comments can be seen as an indication of pre-service teachers' difficulties to interpret students' ideas correctly, which has also been reported in prior research on pre-service teachers in a content-based context (Talanquer et al., 2015). The recent findings add new knowledge to the existing findings from content-based contexts in that they transfer already known difficulties of in-service and pre-service teachers to the field of inquiry related competences. The distribution of feedback comments was moreover analysed with respect to the quality of the student artefact. Looking at the prescriptive as well as the descriptive or evaluative comments, most refer to the poor artefact. This can be interpreted as reflecting the amount of support that is needed. However, it has to be kept in mind that the quality of the best artefact was also only slightly better than average, i.e. it still offered considerable room for improvement. Interestingly, most of the incorrect comments are related to the best artefact. In connection with the low judgement accuracy and the observed differences between bachelor and master students who could be expected to have different levels of PCK, these findings can again be interpreted as an indication of the difficulties pre-service teachers face in the interpretation of student work in the field of inquiry related competences. One possible reason contributing to these difficulties could be that pre-service teachers do not only lack formative assessment practices but also the necessary content knowledge and inquiry related competences. This hypothesis is supported by the findings presented in Table 3 as well as by the qualitative analysis of the feedback quality indicating pre-service teachers' problems with the identification of the different variables. Moreover, it is in line with the observed relation to the quality of the student work. If the pre-service teachers' own knowledge with respect to the CVS is insufficient, they might probably still be able to recognise a poor performance, but have problems judging – and responding to – student work of higher quality. This conclusion can be supported by theoretical considerations that reflect upon the development of teachers' professional knowledge. One can assume that CK is a necessary and sufficient condition for the development of PCK. This assumption is partly supported by results from empirical studies that report moderate correlations between CK and PCK for pre-service teachers (Großschedl, Mahler, Kleickmann, & Harms, 2014; Kleickmann et al., 2014). Furthermore, qualitative evidences relating teachers' interpretations of student learning to their CK support this assumption (Cowie & Bell, 1999; Halim & Meerah, 2002).

In summary, the results illustrate that judging student work and providing potentially effective feedback to support learning are challenging practices for pre-service teachers indicating a need to support the development of these practices in teacher education programmes. A new finding is that these difficulties could also be revealed in the field of inquiry related competences. Based on a literature review, Gotwals and Birmingham (2016) conclude that scaffolding pre-service teachers' formative assessment practices is an important, yet often underdeveloped, area of initial teacher preparation (Buck et al., 2010). Consequently, teachers often develop their formative assessment practices not until they start teaching at school (Hascher, 2008). The presented exploratory study stresses the complexity of formative assessment practices and highlights possible aspects that should be addressed when supporting pre-service teachers' formative assessment practices.

Limitations

The design of the study inevitably results in some limitations that mainly concern the generalisability of the presented findings and thus have to be taken into account when interpreting the results. First, the present study was realised with a non-representative, convenience sample that was rather small in size. Second, it was carried out with a specific focus, namely the CVS that is only one competence (or one part of a competence) in science. Third, it was conducted under conditions of reduced complexity and not in class with, e.g. teaching plans, video recordings and written reflections. Moreover, the participants did not know the students whose artefacts they were assessing which goes hand in hand with the limitation that the effect of the given feedback on students' learning could not be tested. Fourth, the study included no control of pre-service teachers' own knowledge with respect to the CVS, i.e. it was not assessed whether they knew how a good experimental design would look like.

The results thus have to be interpreted in the context of the specific setting. They can provide indications of pre-service teachers' competences and challenges with respect to the accuracy of their judgements and the quality of their feedback but it is not possible to conclude pre-service teachers' general formative assessment practices from the measured data. Another limitation concerns the validity of the used analyses tools. The coding tools used to analyse the judgement accuracy and the quality of the feedback had not been validated against established instruments (due to a lack of suitable, already existing instruments).

Implications for further research

Based on the reported findings, two implications for further research can be concluded: There is a need (1) for further analyses of teachers' subject-specific professional knowledge needed for formative assessment practices and (2) for an empirical analysis of approaches and methods that support the development and use of this knowledge in teaching situations (cf. Blömeke, Gustafsson, & Shavelson, 2015).

One important aspect regarding the first implication is that research regarding formative assessment practices often tends to focus on the practices themselves and overlooks the importance of the teachers' understanding of students' knowledge development. This has been summarised by Coffey, Hammer, Levin, and Grant (2011) as 'the missing disciplinary substance of formative assessment' (p. 1109). Improving teachers' professional competence requires a coherence between scientific concepts, formative assessment practices and consequences for teaching and learning (Dolin et al., 2018; Furtak et al., 2016), i.e. a focus more on scientific concepts and inquiry related competencies than on the assessment practices themselves. By concentrating on the specific topic of the CVS in science and pre-service teachers' formative assessment practices related to this topic, the presented research contributes to a better understanding of the disciplinary substance of these practices in the field of science. Obviously, pre-service teachers themselves lack the subject-specific or domain specific knowledge that is necessary for judgement accuracy and feedback. Prospective research should focus on the normative definition and empirical analysis of pre-service science teachers' knowledge that they need, i.e. for the identification of students' ideas. It is assumed that effective formative assessment practices require profound knowledge, i.e. of how a good performance of subject-specific learning goals looks like or does not look like (Heritage et al., 2009).

A second implication for further research is the development and evaluation of approaches that support the development of pre-service teachers' knowledge and its use with a special focus on future learning. The driving question here is *In what ways can pre-service teachers be supported in using the results of their formative assessment practices in order to adapt learning processes and their teaching?* One challenge is the gap between theory and practice. Having pre-service teachers observe and reflect on teaching videos and students' ideas might be one way of supporting them in bridging the theory that they learn in methods courses and their practices in the field (van Es & Sherin, 2006) and thus helping them to develop the practice of identifying students' ideas. In addition, pre-service teachers need support when interpreting these ideas and especially when interpreting vague nuances expressed by these ideas. Here, learning progressions could be a promising way since they could help pre-service teachers to focus on the relevant aspects of students' thinking (Furtak et al., 2016; von Aufschnaiter & Alonzo, 2018).

Notes

1. The student who wrote the best artefact also identified the dependent and the independent variable. The student who wrote the average artefact did not give this information.
2. All the example comments have been translated from German by the authors.
3. One comment can refer to more than one characteristic of a written investigation plan. Therefore, the aggregated percentages don't add up to 100%.

Acknowledgements:

We acknowledge the support by Hilda Scheuermann, Wiebke Albrecht, Jes-Matthias Stahl and Josephine Kaak when realising the study and analysing the data. Furthermore, we thank all pre-service teachers for taking part.

Disclosure statement

No potential conflict of interest was reported by the authors.

Funding

This work was supported by the European Union's Seventh Framework Programme (FP7/2007–2013) [grant number 321428].

ORCID

Mathias Ropohl http://orcid.org/0000-0002-5932-1453
Silke Rönnebeck http://orcid.org/0000-0003-1638-9543

References

Anders, Y., Kunter, M., Brunner, M., Krauss, S., & Baumert, J. (2010). Diagnostische Fähigkeiten von Mathematiklehrkräften und ihre Auswirkungen auf die Leistungen ihrer Schülerinnen und Schüler [Mathematics teachers' diagnostic skills and their impact on students' achievements]. *Psychologie in Erziehung und Unterricht*, *57*, 175–193.

Arnold, J. C., Kremer, K., & Mayer, J. (2014). Understanding students' experiments – what kind of support do they need in inquiry tasks? *International Journal of Science Education*, *36*(16), 2719–2749.

Bauer, J., Diercks, U., Rösler, L., Möller, J., & Prenzel, M. (2012). Lehramtsstudium in Deutschland: Wie groß ist die strukturelle Vielfalt? [teacher education in Germany: How heterogeneous are study programs?]. *Unterrichtswissenschaft*, *40*(2), 101–120.

Baumert, J., & Kunter, M. (2013). The COACTIV model of teachers' professional competence. In M. Kunter, J. Baumert, W. Blum, U. Klusmann, S. Krauss & M. Neubrand (Eds.), *Cognitive activation in the mathematics classroom and professional competence of teachers. Results from the COACTIV project* (pp. 25–48). New York, NY: Springer.

Bell, B., & Cowie, B. (2001). The characteristics of formative assessment in science education. *Science Education*, *85*(5), 536–553.

Black, P., & Wiliam, D. (1998). Assessment and classroom learning. *Assessment in Education: Principles, Policy & Practice*, *5*(1), 7–74.

Blömeke, S., Gustafsson, J.-E., & Shavelson, R. J. (2015). Beyond dichotomies: Competence viewed as a continuum. *Zeitschrift für Psychologie*, *223*(1), 3–13.

Brookhart, S. M. (2008). *How to give effective feedback to your students*. Alexandria, VA: Association for Supervision and Curriculum Development (ASCD).

Buck, G. A., Trauth-Nare, A., & Kaftan, J. (2010). Making formative assessment discernable to pre-service teachers of science. *Journal of Research in Science Teaching*, *47*(4), 402–421.

Chen, Z., & Klahr, D. (1999). All other things being equal: Acquisition and transfer of the control of variables strategy. *Child Development, 70*(5), 1098–1120.

Cizek, G. J. (2010). An introduction to formative assessment: History, characteristics, and challenges. In H. L. Andrade, & G. J. Cizek (Eds.), *Handbook of formative assessment* (pp. 3–17). New York, NY: Routledge.

Coffey, J. E., Hammer, D., Levin, D. M., & Grant, T. (2011). The missing disciplinary substance of formative assessment. *Journal of Research in Science Teaching, 48*(10), 1109–1136.

Cowie , B. (2012). Focusing on the Classroom: Assessment for Learning. In B. Fraser, K. Tobin, & C. McRobbie (Eds.), *Second International Handbook of Science Education* (pp. 679–690). Dordrecht: Springer.

Cowie, B., & Bell, B. (1999). A model of formative assessment in science education. *Assessment in Education: Principles, Policy & Practice, 6*(1), 101–116.

Dolin, J., Black, P., Harlen, W., & Tiberghien, A. (2018). Exploring relations between formative and summative assessment. In J. Dolin, & R. Evans (Eds.), *Transforming assessment through an interplay between practice, research and policy* (pp. 53–80). Cham: Springer.

Furtak, E. M., Kiemer, K., Circi, R. K., Swanson, R., de León, V., Morrison, D., & Heredia, S. C. (2016). Teachers' formative assessment abilities and their relationship to student learning: Findings from a four-year intervention study. *Instructional Science, 44*(3), 267–291.

Geller, C., Neumann, K., & Fischer, H. E. (2014). A deeper look inside teaching scripts: Learning process orientations in Finland, Germany and Switzerland. In H. E. Fischer, K. Neumann, P. Labudde, & J. Viiri (Eds.), *Quality of instruction in physics: Comparing Finland, Germany and Switzerland* (pp. 81–92). Münster: Waxmann.

Gess-Newsome, J. (2015). A model of teacher professional knowledge and skill including PCK: Results of the thinking from the PCK summit. In A. Berry, P. Friedrichsen, & J. Loughran (Eds.), *Re-examining pedagogical content knowledge in science education* (pp. 28–42). London: Routledge Press.

Gotwals, A. W., & Birmingham, D. (2016). Eliciting, identifying, interpreting, and responding to students' ideas: Teacher candidates' growth in formative assessment practices. *Research in Science Education, 46*(3), 365–388.

Großschedl, J., Mahler, D., Kleickmann, T., & Harms, U. (2014). Content-related knowledge of biology teachers from secondary schools: Structure and learning opportunities. *International Journal of Science Education, 36*(14), 2335–2366.

Hadfield, M. (1995). Das Kupfer-problem [The copper problem]. *ChemKon, 2*(3), 103–106.

Halim, L., & Meerah, S. M. (2002). Science trainee teachers' pedagogical content knowledge and its influence on physics teaching. *Research in Science & Technological Education, 20*(2), 215–225.

Hascher, T. (2008). Diagnostische Kompetenzen im Lehrberuf [Diagnostic competence in the teaching profession]. In C. Kraler, & M. Schratz (Eds.), *Wissen erwerben, Kompetenzen entwickeln. Modelle zur kompetenzorientierten Lehrerbildung* (pp. 71–86). Münster: Waxmann.

Hattie, J., & Timperley, H. (2007). The power of feedback. *Review of Educational Research, 77*(1), 81–112.

Haug, B. S., & Ødegaard, M. (2015). Formative assessment and teachers' sensitivity to student responses. *International Journal of Science Education, 37*(4), 629–654.

Heritage, M., Kim, J., Vendlinski, T., & Herman, J. (2009). From evidence to action: A seamless process in formative assessment? *Educational Measurement: Issues and Practice, 28*(3), 24–31.

Herman, J. L., Osmundson, E., & Silver, D. (2010). *Capturing quality in formative assessment practice: Measurement challenges* (CRESST Report 770). Los Angeles, CA: University of California, National Center for Research on Evaluation, Standards, and Student Testing (CRESST).

Kingston, N., & Nash, B. (2011). Formative assessment: A meta-analysis and a call for research. *Educational Measurement: Issues and Practice, 30*(4), 28–37.

Kleickmann, T., Großschedl, J., Harms, U., Heinze, A., Herzog, S., Hohenstein, F., … Zimmermann, F. (2014). Professionswissen von Lehramtsstudierenden der mathematisch-naturwissenschaftlichen Fächer – Testentwicklung im Rahmen des Projekts KiL [Professional knowledge of

preservice mathematics and science teachers – test construction in the project KiL]. *Unterrichtswissenschaft, 42*(3), 280–288.
Kluger, A. N., & DeNisi, A. (1996). The effects of feedback interventions on performance: A historical review, a meta-analysis, and a preliminary feedback intervention theory. *Psychological Bulletin, 119*(2), 254–284.
KMK. (2005). *Bildungsstandards im Fach Chemie für den Mittleren Schulabschluss [educational standards for chemistry in lower secondary school].* München: Wolters Kluwer.
Landis, J. R., & Koch, G. G. (1977). The measurement of observer agreement for categorical data. *Biometrics, 33*(1), 159–174.
Millar, R. (2017). Using assessment materials to stimulate improvements in teaching and learning. In K. Hahl, K. Juuti, J. Lampiselka, A. Uitto, & J. Lavonen (Eds.), *Cognitive and affective aspects in science education research. Contributions from science education research, Vol. 3* (pp. 31–40). Cham: Springer.
Neumann, K., Härtig, H., Harms, U., & Parchmann, I. (2017). Science teacher preparation in Germany. In J. Pedersen, T. Isozaki, & T. Hirano (Eds.), *Science teacher education: A multi-country comparison* (pp. 29–52). Charlotte, NC: Information Age Publishing.
NGSS Lead States. (2013). *Next generation science standards: For states, by states.* Washington, DC: The National Academies Press.
Nussbaum, J. (1981). Towards the diagnosis by science teachers of pupils' misconceptions: An exercise with student teachers. *European Journal of Science Education, 3*(2), 159–169.
Panadero, E., & Jonson, A. (2013). The use of scoring rubrics for formative assessment purposes revisited: A review. *Educational Research Review, 9*, 129–144.
Park, S., & Oliver, J. S. (2008). Revisiting the conceptualisation of pedagogical content knowledge (PCK): PCK as a conceptual tool to understand teachers as professionals. *Research in Science Education, 38*(3), 261–284.
Rönnebeck, S., Bernholt, S., & Ropohl, M. (2016). Searching for a common ground – A literature review of empirical research on scientific inquiry activities. *Studies in Science Education, 52*(2), 161–197.
Ross, J. A. (1988). Controlling variables: A meta-analysis of training studies. *Review of Educational Research, 58*(4), 405–437.
Ruiz-Primo, M. A., & Li, M. (2013). Analyzing teachers' feedback practices in response to students' work in science classrooms. *Applied Measurement in Education, 26*(3), 163–175.
Schiepe-Tiska, A., Schmidtner, S., Müller, K., Heine, J.-H., Neumann, K., & Lüdtke, O. (2016). Naturwissenschaftlicher Unterricht in Deutschland in PISA 2015 im internationalen Vergleich [Teaching and learning of science in Germany compared to other countries in PISA 2015]. In K. Reiss, C. Sälzer, A. Schiepe-Tiska, E. Klieme, & O. Köller (Eds.), *PISA 2015, Eine Studie zwischen Kontinuität und Innovation* (pp. 133–175). Münster: Waxmann.
Schrader, F.-W. (2013). Diagnostische Kompetenz von Lehrpersonen [Diagnostic competence of teachers]. *Beiträge zur Lehrerbildung, 31*(2), 154–165.
Schwichow, M. G., Croker, S., Zimmerman, C., Höffler, T., & Härtig, H. (2016). Teaching the control-of-variables strategy: A meta-analysis. *Developmental Review, 39*, 37–63.
Schwichow, M., Zimmerman, C., Croker, S., & Härtig, H. (2016). What students learn from hands-on activities. *Journal of Research in Science Teaching, 53*(7), 980–1002.
Seidel, T., & Shavelson, R. J. (2007). Teaching effectiveness research in the past decade: The role of theory and research design in disentangling meta-analysis results. *Review of Educational Research, 77*(4), 454–499.
Siler, S. A., & Klahr, D. (2012). Detecting, classifying, and remediating: Children's explicit and implicit misconceptions about experimental design. In R. W. Proctor, & E. J. Capaldi (Eds.), *Psychology of science. Implicit and explicit processes* (pp. 137–180). New York: Oxford University Press.
Talanquer, V., Bolger, M., & Tomanek, D. (2015). Exploring prospective teachers' assessment practices: Noticing and interpreting student understanding in the assessment of written work. *Journal of Research in Science Teaching, 52*(5), 585–609.

van Es, E. A., & Sherin, M. G. (2006). How different video club designs support teachers in "learning to notice". *Journal of Computing in Teacher Education, 22*(4), 125–135.
Vieluf, S., Kaplan, D., Klieme, E., & Bayer, S. (2012). *Teaching practices and pedagogical innovations: Evidence from TALIS*. Paris: OECD Publishing.
von Aufschnaiter, C., & Alonzo, A. C. (2018). Foundations of formative assessment: Introducing a learning progression to guide preservice physics teachers' video-based interpretation of student thinking. *Applied Measurement in Education, 31*(2), 113–127.

Analysing formal formative assessment activities in the context of inquiry at primary and upper secondary school in Switzerland

Regula Grob, Monika Holmeier and Peter Labudde

ABSTRACT

In this study, 34 trials of formal formative assessment activities in the context of inquiry-based science education (IBSE) from Swiss primary and upper secondary school teachers will be analysed. The aim will be to explore the formal formative assessment activities implemented in IBSE at different school levels.

The trials (14 from primary school, 20 from upper secondary school) are documented with teaching plans and teaching materials as well as with transcripts from individual interviews. For the analysis, the trials were coded with a deductively developed coding scheme involving categories specifying the inquiry characteristics of the trials as well as the formative assessment characteristics.

Overall, the empirical results suggest that at primary school level, more extensive formative assessment practices are possible, such as the combination of several methods within the context of one inquiry unit. At upper secondary school level, formative assessment practices are more straightforward and closely linked to summative assessment practices. These differences will be used to develop characteristic features of formative assessment practices for the two school levels.

Introduction and research question

The effect of formative assessment (FA) on student learning is high (Black & Wiliam, 1998; Hattie, 2009). This 'highlights the importance of firmly embedding formative assessment within the broader evaluation and assessment framework and the need to support teachers' capacity and professionalism in formative assessment' (OECD, 2013, p. 145) in international educational policy. However, the success of such measures depends on their implementation in teaching practice (Black, 1993; Black & Wiliam, 1998; Stiggins, Griswold, & Wikelund, 1989). FA is employed in classrooms, so its quality depends on the strategies individual teachers use to find evidence for student learning and to build further instruction and learning on this evidence (Bell & Cowie, 2001; Heritage, 2010; Herman, Osmundson, & Silver, 2010; Ruiz-Primo, Furtak, Ayala, Yin, & Shavelson, 2010).

In Switzerland, a new curriculum *Lehrplan 21* (curriculum 21) is currently being introduced. Reflecting international developments, the importance of assessment in general but FA in particular, is stressed in this national document (D-EDK, 2014). In daily teaching, teachers have a high level of autonomy in this country. FA activities have been known and used in the classrooms for the last few decades, but the variability between teachers and between school levels has been high (Vögeli - Mantovani, 1999).

In this context, the study can be considered as a first step towards understanding how the transfer of FA from the level of national and international educational policy to daily teaching practice could be supported. The focus of the paper will therefore be on how teachers at Swiss primary- and upper secondary school utilise formal formative assessment methods. More tangibly, the research question explored will be: how and how often do teachers at Swiss primary- and upper secondary school implement formal formative assessment methods in their IBSE-teaching? This will include the way the expectations are articulated and shared with the students, the data used for diagnosis of student level, the formative assessment methods used, the competences assessed, as well as the way the feedback is used for further learning.

As a context for the study, inquiry-based science education was chosen because of its growing importance in science curricula and teaching practice. However, this context is not an easy one for Swiss teachers (Börlin, 2012; Smit, Weitzel, Blank, Rietz, Tardent, & Robin, 2017; Grob, 2017). Examples of good practice for formative assessment in the context of inquiry-based science education can be found – in broad outlines - in Black & Harrison, 2004, but for the empirical part of this paper, it appeared necessary to define characteristic features to distinguish inquiry-based science units from others. More details on these features will follow in the first section of the theoretical part just below.

Inquiry-based science education as the context of this study

Inquiry-based science education (IBSE) subsumes many different approaches to teaching and learning (Furtak, Seidel, Iverson, & Briggs, 2012). However, three characteristic features to operationalise the term can be deduced from the many definitions in the literature: firstly, IBSE is described as a number of student activities that resemble research: observing and investigating phenomena, shaping questions and hypotheses, collecting and interpreting data, reasoning and communicating (e.g. Bybee, 1997; Rönnebeck, Bernholt, & Ropohl, 2016). Secondly, enhanced student competences are considered to be one of the goals of IBSE (e.g. Abd El Khalick et al., 2004). Students' inquiry competences, like observing and investigating phenomena, are supposed to be fostered by IBSE. Finally, inquiry activities are open to some degree, they leave some decisions to the individual students rather than pre-defining all aspects and procedures (e.g. Priemer, 2011). These three characteristic features will be described in more detail below.

Inquiry activities and procedural character of inquiries

IBSE has often been conceptualised as a set of student activities. Often, in practice, these activities do not follow each other one by one. Rather, inquiry processes contain loops back to earlier steps and revisions of certain activities (Artigue & Baptist, 2012). Also, an inquiry unit does not need to stop after the conclusion, which is typically the last step. Instead, a conclusion could initiate a new inquiry process with a subsequent question (e.g. White & Frederiksen, 1998).

Bell, Urhahn, Schanze, and Ploetzner (2010) derived a tangible set of activities from the international literature on IBSE which they call the 'nine main processes of inquiry learning'. They include: orienting and asking questions, hypothesis generation, investigation, analysis and interpretation, modelling, conclusion and evaluation, communication, and prediction (Bell et al., 2010).

Domain-specific and transversal competences ascribed to inquiry

As stated, a second aspect ascribed in IBSE is competence orientation. The term competence is described as 'context-specific cognitive dispositions that are acquired by learning and needed to successfully cope with certain situations or tasks in specific domains' (Hartig, Klieme, & Leutner, 2008, p. 9). Most EU-projects on inquiry (e.g. Mind the Gap, S-TEAM, ESTABLISH and Fibonacci) describe IBSE as a set of activities and also consider these activities to be inquiry competences that can be acquired by students. This paper will follow the same understanding. Consequently, the set of activities defined in Bell et al. (2010), introduced above, will also be considered as domain-specific inquiry competences: orienting and asking scientific questions, hypothesis generation based on scientific knowledge, planning an inquiry, investigating a scientific problem, analysing and interpreting the results, modelling the problem, concluding and evaluating the inquiry, communicating about the methodology and the results, and making predictions for further research.

Apart from these domain-specific competences, a number of authors ascribe transversal competences to IBSE too (e.g. Zachos, Hick, Doane, & Sargent, 2000). Transversal competences, also known as 'Cross Curricular Competences C^3', are transversal to the structure of scholarly disciplines and therefore cannot be assigned to a specific subject (Grob & Maag Merki, 2001). The transversal competences defined in the DeSeCo project (OECD, 2005) are broad enough to cover the different understandings of transversal competences as employed in daily teaching practice in Switzerland. It will therefore form the basis for the conceptualisation of transversal competences in this paper. The competences are classified in three groups (OECD, 2005): firstly, the interactive usage of tools such as language, knowledge and technology; secondly, interacting in heterogeneous groups; and thirdly, acting autonomously.

Openness of student activities in the context of scientific inquiry

Characteristic to IBSE is the fact that the student activities are not entirely pre-defined but open: they leave some decisions to the students. Priemer (2011) developed a model with different dimensions of openness in science experiments which is applied to inquiry activities in general in this paper:

- Openness in terms of content
- Openness in terms of strategy
- Openness in terms of methods used

- Openness in terms of the number of possible solutions
- Openness in terms of the number of different solution processes.

Formative assessment to enhance inquiry-based science education

In IBSE, with its openness of student activities, the traditional roles of students and teachers change (Kessler & Galvan, 2007): students take an active part in deciding on the questions investigated, on the collection and interpretation of the data gathered, and in the development of explanations from evidence (Euler, 2011). The teachers' role is to initialise, coach and support the inquiry process (Kessler & Galvan, 2007; McLoughlin, Finlayson, & van Kampen, 2012). One approach to such coaching and support of the students in their inquiry learning is FA (Dolin, 2012) as outlined in the following section.

Methods of formal formative assessment

Formative assessment has the aim of assisting student learning and is therefore also called assessment for learning. It involves processes of 'seeking and interpreting evidence for use by learners and their teachers to decide where the learners are in their learning and where they need to go and how best to get there' (ARG, 2002, p. 2).

One way to describe FA activities is by their degree of formality (Shavelson et al., 2008): there is a range from informal to formal activities which can be distinguished by the planning involved, the type of student data worked with, as well as the nature of the feedback to the learner (Cowie & Bell, 1999). In informal FA, an example being assessment on-the-fly, the situations that occur in a lesson are spontaneously used for FA. Formal FA, on the other hand, is based on activities which have been previously planned for this specific purpose. The process in FA follows four steps: the expectations are articulated in the form of criteria, a diagnosis of the student level is made based on a particular type of student data, the student is provided with feedback on his or her learning, and the student is given the opportunity to use this feedback (e.g. Paris & Paris, 2001).

Three of the formal formative assessment methods that are most frequently described in the literature are written teacher assessment, peer-assessment, and self-assessment. Written teacher assessment can be based on criteria lists (e.g. Burke, 2006; Moskal, 2003) and on open comments (e.g. Black & Harrison, 2004). Often, the varieties are combined: teachers use assessment templates which contain both a list of criteria for formative purposes and space for open comments.

Peer-assessment is based on the idea of 'activating students as instructional resources for one another' (Leahy, Lyon, Thompson, & Wiliam, 2005, p. 21): students assess other students' work as assessors and receive their peers' feedback as assessees. Consequently, the benefit of peer-assessment is that students learn to identify strengths and weaknesses of their peers' work and also receive suggestions on their own work. Both processes aid the reflection upon the students' own learning processes (Topping, 2003; Tsivitanidou, Zacharia, & Hovardas, 2011). A variety of tools for peer-assessment, including rating instruments and checklists has been described in the literature (Falchikov, 1991).

Self-assessment 'is a process of formative assessment during which students reflect on the quality of their work, judge the degree to which it reflects explicitly stated goals or criteria, and revise accordingly.' (Andrade, 2010, p. 91). The basic idea of self-assessment is similar to the above-described peer-assessment: students are prompted to reflect upon the quality of their work themselves rather than consider the teacher's judgement (Andrade, 2010).

Four steps for formal formative assessment

As stated above, formal formative assessment generally follows four steps: the expectations are articulated in the form of criteria, a diagnosis of the student level is made based on a particular type of student data, the student is provided with feedback on his or her learning, and the student is given the opportunity to use this feedback (e.g. Paris & Paris, 2001).

Articulating and sharing the expectations of the learning is a prerequisite for the student to demonstrate and the teacher to diagnose, the student level of achievement. Such articulation can manifest itself in different formats: explicit communication by the teacher (e.g. Andrade & Valtcheva, 2009), and elaboration and co-construction with the students (e.g. Black, Harrison, Lee, Marshall, & Wiliam, 2004). Furthermore, the criteria for a particular unit can also be implicitly clear (for example if they are the same for a longer period).

The diagnosis of the student level of achievement could be based on different types of data including written reports, oral conversation with the students, and observation of the students' actions by the teacher. Once the data is gathered, it needs to be analysed and interpreted in order to determine where the student is in relation to the learning goals.

The feedback to the learner is offered in order to plan subsequent learning: based on the analysis and interpretation of the student data, feedback to the student on how to proceed is provided (in the case of student-oriented activities) or subsequent teaching is shaped (in the case of teacher-led activities; Hattie & Timperley, 2007).

In the context of student-oriented activities, the use of the feedback could be a revision of the original piece of work, such as the revision of the planning for an experiment. As a second option, the feedback could also be transferred to a similar, new situation (e.g. Andrade & Valtcheva, 2009; Paris & Paris, 2001). An example for the second option is that students receive feedback on their documentation of an experiment and use this feedback in a subsequent unit, when they document another experiment. The decision on the use of feedback depends on the degree of planning involved, the level of formality of the feedback, the type of data on student learning searched for, and the type of feedback (Furtak & Ruiz-Primo, 2008).

Inquiry and assessment in Switzerland

Compared to other countries, Switzerland has a highly de-centralised educational system. This has two reasons: on the one hand, the country has a strong federal tradition and the individual *cantons* (states) have a high degree of freedom in educational decisions. On the other hand, the teachers have a high level of autonomy and far-reaching responsibilities in their teaching. This autonomy includes, for example, decisions on teaching methods such as IBSE, but also the development of almost all exams and the respective decisions on grades. More details on the context of the study will be provided in the next section.

Current standing of inquiry-based science education in Switzerland

At primary and lower secondary school level, a new curriculum valid for each whole linguistic region has started being implemented. In the science subjects, its basis is a competence model with minimal standards developed in the HarmoS project (HarmoS: Konsortium HarmoS Naturwissenschaften, 2008). Both the competence model and the science curricula foster inquiry approaches (Labudde, Nidegger, Adamina & Gingins, 2007; Labudde, 2007).

At the gymnasium, which represents the upper secondary school level in Switzerland and is attended by 20% of the students, the alignment of the curricula started earlier, with the present curriculum being used since 1997. This curriculum distinguishes between knowledge, skills and attitudes in the curricula (EDK, 1994). The need for IBSE might, to some extent, be indicated in the skills domain.

A limited number of empirical studies explored the classroom situations regarding IBSE at primary (HarmoS: Konsortium HarmoS Naturwissenschaften, 2008), lower (Börlin, 2012; Börlin & Labudde, 2014) and upper secondary school (Labudde, 2000; Widmer Märki, 2011) in Switzerland. The small-scale explorative studies cannot provide a valid overview of the general IBSE practices in Switzerland. Nevertheless, the emerging picture is that indications for inquiry-based educations do exist at all school levels but with ample room for improvement in various aspects. These aspects include: the lack of enough rooms and equipment at primary school (HarmoS: Konsortium HarmoS Naturwissenschaften, 2008), the focus on conducting experiments rather than preparing and post-processing them (Börlin, 2012), the scarce employment of student-centred activities (Börlin & Labudde, 2014), and the rare use of innovative forms of summative assessment (Widmer Märki, 2011).

Only anecdotal evidence was found on the students' inquiry abilities in Switzerland. In the international literature (e.g. Blanchard, Osborne, Wallworks, & Harris, 2013), however, it is reported that there are significant differences between school levels.

Formative assessment traditions in Switzerland

The curriculum for the compulsory school levels which is presently being implemented strengthens the role and importance of FA (D-EDK, 2014). In the curriculum for the Gymnasium level (upper secondary school, attended by about 20% of students), no guidelines on assessment exist. FA activities are included in the newer school books in all subjects, in particular at primary and lower secondary school level.

Little is known about the employment of FA practices in Swiss classrooms. Such practices are generally accepted by teachers (Vögeli - Mantovani, 1999), their use varies greatly between school levels and also at the level of individual teachers (Kronig, 2009). This is possible due to the legal standards which provide the teachers with a high level of autonomy in their teaching and individuality at the level of school units and even classrooms (Husfeld, 2009; Kronig, 2009; Vögeli - Mantovani, 1999). Furthermore, typical assessment practices are dominated by summative goals and some teachers seem to lack abilities to diagnose competences in general and different levels of proficiency in particular, as well as time to employ FA practices (Jundt, 2013; Smit, 2009; Smit & Birri, 2014).

Research question

As stated earlier, the empirical part of the paper will focus on how teachers at Swiss primary- and upper secondary school utilise formal formative assessment methods. More tangibly, the research question explored will be: how do teachers at Swiss primary- and upper secondary school implement formal formative assessment methods in their IBSE-teaching? This research question will include the way the expectations are articulated and shared with the students (sub-question 1), what type of data are used how often for diagnosis of student level (sub-question 2), what formative assessment methods are used (sub-question 3), what competences are assessed (sub-question 4), as well as what way the feedback is used for further learning (sub-question 5).

Methods

In order to investigate the research questions stated above, a collaboration with $N = 20$ science teachers from primary and from upper secondary school was established. The teachers were asked to trial formative assessment in the context of IBSE. The first set of data collected consisted of a written report form on the context of the IBSE-units and on the procedure of the FA including the teaching plans and their teaching materials. These were coded deductively and proceeded with qualitative content analysis. For triangulation purposes, assessed student works and the respective feedback as well as observation notes from classroom visits were collected.

The second set of data consisted of interviews with a sub-group of the teachers in which they justified their choices of FA methods and competences assessed. The rationale for this mixed-methods design was to explore their choice of methods and competences for assessment in more detail, linking to the complementarity purpose (Greene, Caracelli, & Graham, 1989; Schoonenboom & Burke Johnson, 2017).

Details will be provided in the following sections.

Trials

In an introductory meeting with the participating teachers, the conceptual understanding of FA in the study was presented and discussed amongst the teachers. The FA methods used in the study were introduced with concrete examples of inquiry-based science units.

Afterwards, the teachers were asked to select one of the FA methods introduced and to implement it in their classrooms in an IBSE unit in order to assess the domain-specific inquiry competences (introduced in the theory part of the paper) at any time during the semester. No restrictions on the number of competences assessed or on the lengths of these units were made. Ideally, this process should have produced 60 implementations (20 teachers implementing 1 FA method per semester during 3 semesters).

Sample

In the sample of $N = 20$ science teachers, there were 9 primary and 11 upper secondary school teachers. These two school levels were selected because the respective teachers are socialised differently. Whereas primary school teachers in Switzerland have a pedagogically-oriented background, upper secondary school teachers are much more subject-oriented. So, the ratio of the study was not to compare school levels but to explore a wide range of approaches to teaching and learning. This should be facilitated by including two different school levels.

At primary school, the subject taught is called integrated science whereas at upper secondary school, the teachers taught biology, chemistry, or physics. All teachers in the sample had a teaching degree for their respective school levels. 6 out of the 9 primary school teachers were female (66.7%); 5 out of the 11 upper secondary school teachers were female (45.5%). The Table 1 below provides an overview of the participants' teaching experience.

Data collection

Written report form for teachers

The teachers were asked to fill out a report form of four pages on every implementation of an IBSE unit with FA right after the trial. This form contained closed questions on the assessment method and on the competences assessed as well as open questions on the context of the unit and on the procedure of the FA. The form also included a template of a teaching plan which had to be filled out. Furthermore, the teachers were asked to provide their teaching materials, such as worksheets.

Individual interviews with teachers

A sub-group of teachers were asked to participate in individual, semi-structured interviews. The respective teachers were selected so that they reflect the two different school levels, subjects taught, and amount of teaching experience. These interviews were conducted no later than 10 days after the implementations of FA and lasted around 40 min. The questions covered aspects of the implementation like the decisions during the preparation, issues in the conducting the assessments, as well as the teachers' evaluations of the implementations.

The resulting data consisted of 5 interviews with primary school teachers and 7 interviews with upper secondary school teachers. The interviews were audiotaped in dialect and afterwards transcribed in standard German. For this paper, selected quotes were translated into English by the authors.

Selection of trials for analysis

20 teachers trialled FA methods in their IBSE units. With one trial per semester and the collaboration lasting for three semesters, this resulted in 54 trials (not 3*20 = 60 trials because some teachers left the project earlier). These units lasted between 2 and 6 lessons and covered topics such as investigation of forest soils or experimentations on buoyancy at primary school and inquiries on electric circuits, sound waves or

Table 1. Teaching experience of the participants.

	0–4 years	5–10 years	11–20 years	>20 years
Teachers from primary school	3 teachers (33.3%)	2 teachers (22.2%)	2 teachers (22.2%)	2 teachers (22.2%)
Teachers from upper secondary school	3 teachers (27.3%)	3 teachers (27.3%)	3 teachers (27.3%)	2 teachers (18.1%)

photosynthesis at upper secondary school. These trials were triaged according to two criteria: the trials had to take place in the context of IBSE and they had to contain formal formative assessment activities.

The first criterion, whether or not the trials were conducted in an IBSE unit, was evaluated based on three indicators: the units had to be open in at least one dimension, they had to include inquiry activities, and the FA activities had to be used to assess domain-specific or transversal competences ascribed to IBSE. For further descriptions of these three indicators see the theory part of this paper.

The second criterion, whether or not the trials contained formal formative assessment activities, was evaluated based on five indicators: the expectations of the learning had to be clear to both the teacher and the students from the beginning of the unit, timing and procedure of diagnosis had to be planned and clear to both the students and the teacher, students had to receive the results of that diagnosis as individual feedback on their piece of work, students needed to have the opportunity to use the feedback, and finally, the FA activities had to be assignable to a formal FA method. In order to fulfil the criterion, all five indicators had to be fulfilled. For further descriptions of these five indicators see the theory part of the paper.

All 54 trials produced in the course of the study were classified depending on whether they fulfilled the criteria with all their indicators and were documented well enough to be analysed. This was true for 14 trials from primary school and 20 trials from upper secondary school, the respective data will be presented in the results part.

Data analysis

Deductive coding (Mayring, 2010) was applied to the written report forms, in which the teachers explained their implementations and laid out their lesson plans (based on a template). Furthermore, the teaching materials such as worksheets were also analysed using deductive coding. Table 2 provides an overview of the dimensions of the coding frame.

The reliability of the coding was calculated by double-coding 18% of the data (10 trials out of 54), resulting in an interrater-reliability, measured with Cohen's Kappa, of $\kappa = 0.83$ (Landis & Koch, 1977). In order to check validity, the results were triangulated with the assessed student works and the respective feedback from 9 implementations as well as with the observation notes from 6 classroom visits.

The transcripts of the individual interviews were inductively coded (Mayring, 2010).

Results

Communication of the criteria

During the course of the IBSE units, the students were familiarised with the assessment criteria for the FA activities. In most trials, the teachers pre-planned these criteria and handed them out to the students. This was the case for 11 of the 14 primary school trials and 16 of the 20 upper secondary school trials. Some of these teachers provided the criteria in the form of questions, others in the form of lists of criteria. Two of the primary school teachers did not provide pre-defined assessment criteria but elaborated

Table 2. Coding frame.

Dimensions (compare to theory part of the paper)	Categories
Communication of criteria (Andrade & Valtcheva, 2009; Black et al., 2004; Ruiz-Primo et al., 2010)	Explicit communication by teacher Elaboration of criteria with students Implicit communication
Data sources for diagnosis (Ruiz-Primo et al., 2010)	Written student data 3-dimensional model/3-dimensional product Oral student data Observational data
Assessment methods (Moskal, 2003 and Black & Harrison, 2004 on written teacher assessment; Topping, 2003 on peer-assessment; Andrade, 2010 on self-assessment)	Written teacher assessment Peer-assessment Self-assessment
Competences assessed (Bell et al., 2010; OECD, 2005)	Domain-specific competences • Orienting and asking scientific questions • Hypothesis generation based on previous scientific knowledge • Planning an investigation • Scientific investigation • Analysis and interpretation of the results • Modelling the problem • Conclusion and evaluation of the investigation • Communication about the methodology and the results • Prediction for further research Transversal competences • Using tools interactively • Interacting in heterogeneous groups • Acting autonomously
Means of engaging with the feedback (Andrade & Valtcheva, 2009; Paris & Paris, 2001)	Revision of original student data Transfer to new task

them together with the students as part of the IBSE unit. In the remaining 5 trials, the assessment criteria were obvious from the context and were not explicitly introduced in this very IBSE unit (e.g. if the focus of all the lab lessons of a whole semester was on documenting experiments or inquiries, the respective criteria were not necessarily introduced in the very unit with FA).

Data sources for diagnosis

According to the literature introduced in the theory part of the paper, the data which the FA was based on was allocated to four categories: written student data such as entries in lab journals, artefacts and models, oral student data, such as student explications and presentations, and finally, observations in which the student's behaviour was observed by the assessor. In the trials analysed in this paper (see Figure 1), written student data and observational data were most common in the primary school trials, whereas written student data was by far the most commonly used source of data in the upper secondary school trials. Artefacts and oral student data were considered more rarely for FA at both school levels.

In some of the trials, more than one source of data was used for the FA. This was the case in 7 out of the 14 primary school trials. At upper secondary school level, this procedure was applied less frequently, most FA activities there were based on one source of student data.

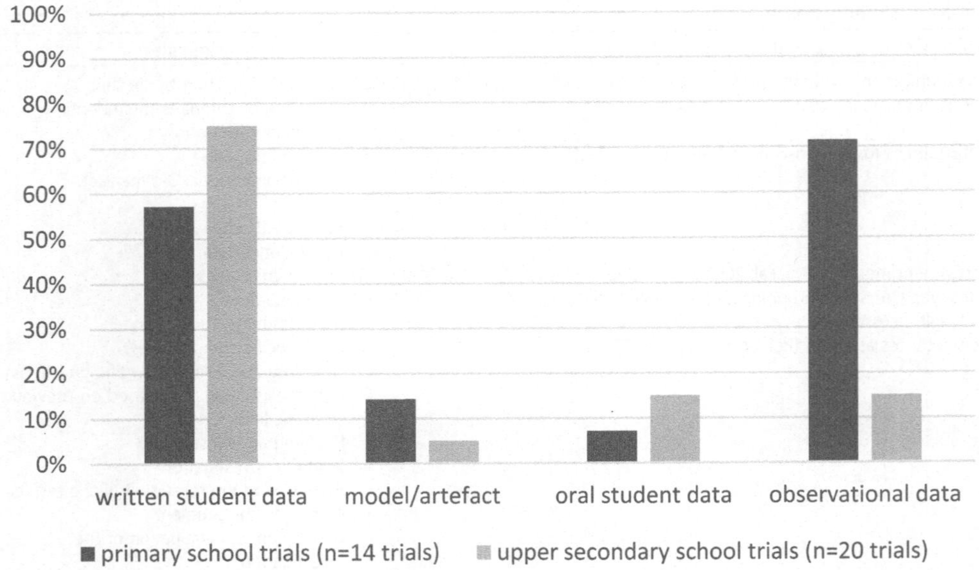

Figure 1. Data sources used for diagnosis. n represents 100%, since some of the trials involved more than one data source the sum of all data equals more than 100%.

Assessment methods

The teachers collaborating in the study embedded at least one FA method in their IBSE units. Some teachers decided to work with more than one FA method in the same unit, resulting in more assessment methods than trials (see Figure 2). In all of the latter cases, the different assessment methods targeted different competences.

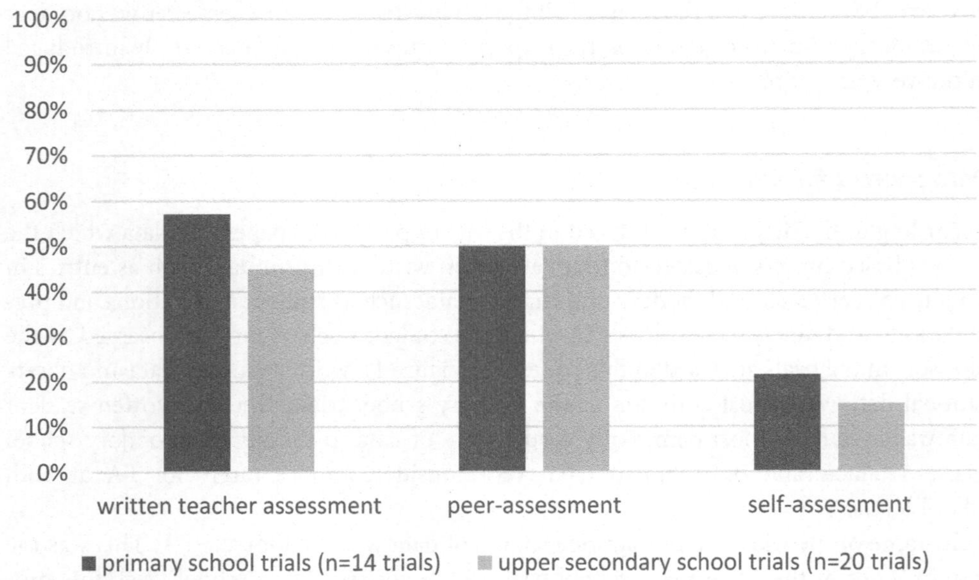

Figure 2. Assessment methods used. n represents 100%, since some of the trials involved more than one assessment method the sum of all data equals more than 100%.

The teachers implemented the formal formative assessment methods in different ways: written teacher assessment was provided in the form of filled-out criteria lists, open comments, in a combination of lists and open comments, and as formative comments on graded reports.

In the peer-assessment trials, the tools mentioned above were limited to criteria lists. An additional distinction was made between individual and group work. Student data could be produced individually or in groups, and the FA could take place individually or in groups. All possible combinations appeared in the study.

In the self-assessment trials, three varieties appeared in the study: the teachers had the students reflect individually by providing reflection sheets with questions (1 primary school trial, 1 upper secondary school trial), reflective discussions amongst groups of students were set up (2 primary school trial, 1 upper secondary school trial), or the reflective discussion took place between the teacher and individual students (1 upper secondary school trial).

The teachers were asked about the reasons for their choice of a particular FA method in the individual interviews. 5 lines of argumentation could be distinguished: firstly, some teachers from both school levels argued that there was no conscious decision but that the FA method '*just emerged*', '*organically grew*', or '*just appeared suitable*'. Secondly, some teachers from both school levels related the choice of the FA method to their self-efficacy, saying things like '*I thought I would be able to manage this*', '*it appeared doable*', or '*I had used the method before and it worked out well*'. Thirdly, some teachers from primary school based their decision on student motivation, giving reasons like '*I thought the students would like this*'. Fourthly, some upper secondary school teachers related their choice to the students' development of reflective skills '*the method will train the students to reflect upon their work*'. Finally, some upper secondary school teachers gave organisational reasons such as the class size or the limits in lesson time.

Competences assessed

The teachers' trials were analysed in terms of what competences they assessed. The results can be found in Figure 3 (domain-specific competences). Most teachers from both school levels included more than one competence in their FA, which is why the number of trials and the number of competences assessed is not the same. Formal aspects like orthography and also content knowledge were included in several trials but are not covered in this analysis.

As can be seen in Figure 3, the majority of domain-specific IBSE competences from Bell et al. (2010) were covered in the study. The frequency of coverage varies greatly, however, 'investigation' and 'communication' were assessed most often. The 'documentation' competence often manifested itself in lab journal entries or presentations of results. Other domain-specific competences were only rarely or never addressed: 'orienting and asking scientific questions', 'hypothesis generation based on scientific knowledge', 'modelling the problem', 'concluding and evaluating the inquiry', and 'making predictions for further research'. When comparing the two school levels, it appears that the primary school trials cover a narrower range of competences than the upper secondary school trials.

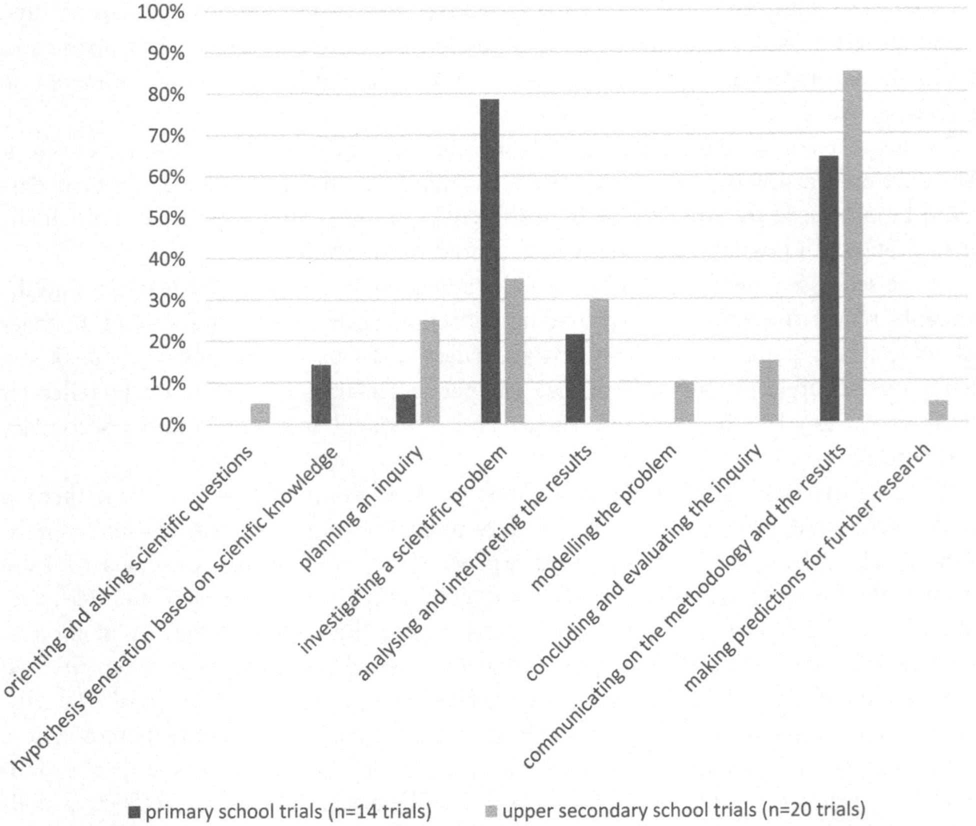

Figure 3. Domain-specific competences assessed. n represents 100%, since some of the trials involved the assessment of more than one competence the sum of all competences equals more than 100%.

The teachers who participated in individual interviews were asked to explain the choice of the competences for FA. Some teachers from both school levels mentioned the relevance of a particular competence when giving reasons for 'investigation' (at primary school), for 'planning an inquiry', 'modelling the problem', 'concluding and evaluating the inquiry' as well as 'communicating about the methodology and on the results' (at upper secondary school). Some of the primary school teachers gave two other explanations: three of them said that the choice of the competence was not a conscious decision but '*just emerged*' or '*appeared suitable*' whereas one other teacher took a resource-based decision, '*found a convincing list of criteria*'. At upper secondary school, two other reasons were offered: the students' abilities in the quote '*had the impression that the students would be able to assess this*' (referring to 'communicating about the methodology and the results') and the relevance of a particular competence for the students' further learning in quotes like '*I think it is important to prepare for university*' (referring to 'communicating about the methodology and the results').

Some of the teachers decided to include transversal competences in their FA activities which were classified into the three categories 'use tools interactively'; 'interact in heterogeneous groups' and 'act autonomously' according to the framework provided in OECD (2005), see theory part on domain-specific and transversel competences ascribed to inquiry. All these transversal competences were assessed in combination with domain specific competences in the same trial, however in some cases with different assessment methods. In one of the primary school trials, for example, the teacher focussed her written assessment on the modelling process whereas the students assessed their peers' abilities to collaborate in a group (classified as 'interact in heterogeneous groups' according to OECD (2005)). Looking at the specifities of the two school levels explored, the vast majority of trials at primary school did involve assessment of one or two transversal competences whereas most of the trials at upper secondary school did not but instead exclusively focussed on domain-specific competences.

Means of engaging with the feedback

The teachers found two ways of ensuring that the students had the possibility of using the feedback they had received: in some trials, original pieces of work could be revised based on the feedback. In other trials, the feedback could be transferred to a subsequent, similar situation. As displayed in Figure 4, the first possibility (revision) was more frequently put into practice in the primary school trials whereas the second possibility (transfer) was more common at upper secondary school. Another relation of the engagement with the feedback was found with the competences assessed: revisions of original pieces of work were more often offered for domain-specific competences whereas transfer to similar tasks occurred more often for transversal competences.

Interpretation of the results in the context of the Swiss educational system

Communication of assessment criteria

The results show that in most trials, the assessment criteria were pre-defined and communicated by the teacher. One possible explanation is that this is the fastest way to get FA processes started and that most teachers decided to spend their lesson time on other aspects of the FA. This could be taken as a sign that the communication of the assessment criteria was not considered the most relevant aspect of the FA process. Another explanation is that some teachers were simply unaware of other options such as elaborating assessment criteria together with students.

Data sources for diagnosis

The student data used for diagnosis of the students' levels of achievement correlates to the competences assessed. In the primary school trials, observational data is appropriate to assess hands-on investigative parts of IBSE (compare with section on competences assessed). Using two sources of student data in one trial could be a sign of the holistic approach to student learning followed by many primary school teachers with their pedagogical socialisation (see theory part of the paper).

In the upper secondary school trials, the use of written student data in order to assess the 'communicating about the methodology and the results' competence appears plausible, too. Widmer Märki (2011) found that upper secondary science teachers focus on the summative purposes of assessment and therefore prefer strong, undeniable evidence like written data rather than observations. In this context, the most frequent choice of

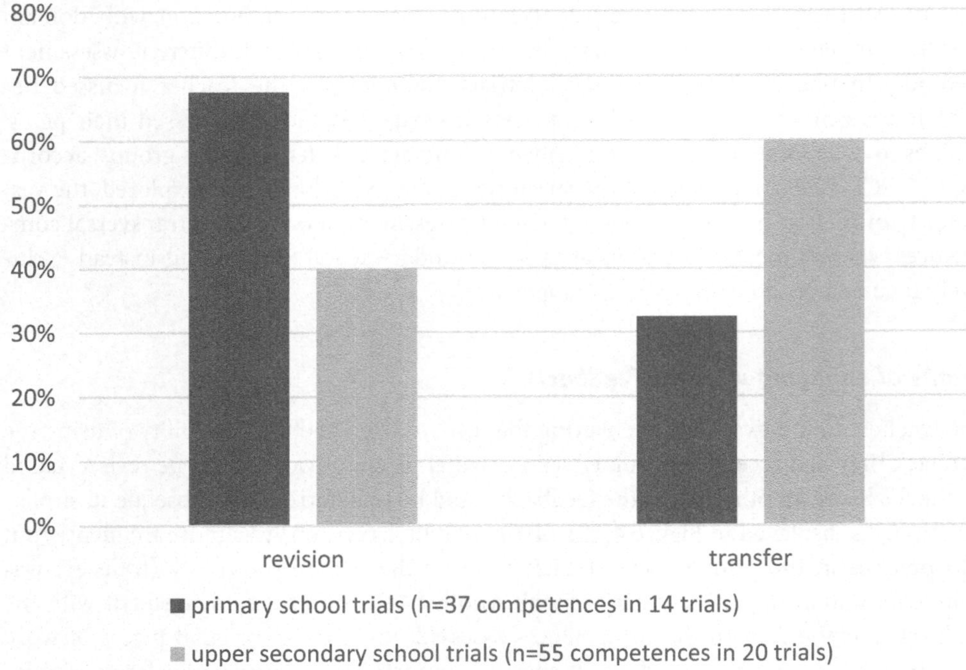

Figure 4. Means of engaging with the feedback received.

student data and competence for FA might reflect the upper secondary school teachers' grading habits.

Assessment methods used

The three formal FA methods were all implemented at both school levels in a variety of enactments. This implies that the FA methods are suitable for the context of IBSE in the Swiss educational context.

Some teachers decided to implement more than one assessment method in one trial, with different assessment methods assessing different competences. This tendency was found by Widmer Märki (2011) too, and she took it as a sign that the respective teachers had some experience or expectation as to which assessment method could combine well with a particular competence. Examples of such considerations could include, for example, what competence could be reflected on by the student him- or herself (self-assessment) or what competence could be well assessed by peers (peer-assessment).

The data in the study shows a slight tendency for domain-specific competences to be assessed more often by the teacher or by the peers whereas transversal competences were frequently reflected upon. An explanation could be that the transversal competences as defined in this study (see theory part of the paper) could be difficult for people other than the students themselves to diagnose.

Finally, the interview data on the teachers' planning of formal FA activities implies that there are two main groups of teachers. For some of them, planning the FA activities was not a process of conscious and explicit choice and reasoning but part of their teaching instinct. Other teachers took their specific situation into account. The topic they wanted to work on, the abilities and the motivation of a particular student group, the teachers' self-efficacy, and also organisational issues. These results can be taken as a sign that to promote the use of formative assessment in classrooms, it is not enough to simply state that FA should be done. Rather, teachers need to experience the benefits of FA and to incorporate it into their teaching without consciously considering it as something 'extra'.

Competences assessed

In the primary school trials, the 'investigating a scientific problem' competence was assessed most often. Three reasons could help to explain this effect: firstly, it is possible that a domain-specific competence which can be directly observed from practical work could be feasible for this school level. Secondly, it might be easier for primary school teachers to develop concrete indicators for diagnosis for a competence reflecting hands-on work compared to other competences such as hypothesis generation. Thirdly, it could also be that the primary school teachers weighted this hands-on part of IBSE as more relevant than other aspects and therefore decided to formatively assess it.

Looking at the upper secondary school trials, the most frequently assessed domain-specific competence was 'communicating about the methodology and on the results'. One possible reason, the vicinity to summative assessment, has already been discussed. Another aspect could be the importance of written communication at upper secondary school level; this competence could be chosen for assessment because students are more proficient in writing than they were at primary school level.

Most teachers included a variety of domain-specific competences and criteria in their trials rather than focussing on one specific aspect. This habit was also found by Widmer Märki (2011) in non-traditional summative assessment (i.e. project presentations, portfolios and similar).

Looking at the interview data and the teachers' reasoning for the assessment of particular domain-specific competences, it is striking that the content of the concrete inquiry unit or the curriculum were never mentioned. Instead, some teachers did not seem to consciously choose a particular competence whereas another's choice was based on their personal conviction as to which competences were important. Again, this could be taken as a sign that for the implementation of FA, it is not enough to just include it in the curriculum. Rather, the teachers need concrete support to incorporate FA into their daily teaching habits which they can employ without much consideration.

Particularly at primary school, many teachers included transversal competences in their trials, even if this was not an explicit part of their task in the study. A possible explanation is that these transversal competences are considered relevant by the teachers (for IBSE or in general), even though they are not part of the usual summative assessment practice in Switzerland. So, FA could be an opportunity to stress the weight of these transversal competences outside the grading routines.

Means of engaging with the feedback

Considerable differences in the use of the feedback were found between the two school levels. At primary school level, revisions of the original piece of work were more often employed. A potential reason is that the transfer itself might be a hard task for a primary school student. Another factor could be that at primary school, the students work at a certain topic or task longer and more thoroughly so that they can be given the time to revise a piece of work.

At upper secondary school, transfer tasks might be more appropriate for the abilities of the students and more feasible because of the time pressure. Also, many of the trials at this school level took place in the context of regularly recurring lab lessons where lab journal entries and lab reports are typically used. Providing FA on one lab report or journal entry and making use of it in the next piece of written work appears feasible.

Across the two school levels, the revision of original pieces of work was more frequent for domain-specific competences whereas transfer to later tasks typically concerned transversal competences. This can be explained with the nature of the competences: the transversal competences are almost exclusively linked to a particular situation such as the behaviour in a particular group or capability to work independently, and the respective feedback could hardly be used for a revision.

Conclusion: formative assessment practices for the school levels explored

The data on the trials in the study shows the breadth of FA activities in the context of IBSE. This variability can be taken as a sign of the Swiss teachers' autonomy in their teaching across all school levels and the adjacent far-reaching responsibilities. The results can also be taken as a sign of how small the influence of educational policy is in terms of the implementation of FA in daily teaching practice. A striking outcome of the interview data is that many teachers did not consciously decide what assessment method to choose or what competences to assess but let their teaching units 'grow organically'.

However, patterns in the results reflecting differences between the school levels could be found and they are summarised in Table 3. These differences in the trials can be explained by the different socialisation: whereas primary school teachers in Switzerland have a pedagogically-oriented background and often aim at holistically educating their children, upper secondary school teachers are much more focussed on teaching subject-specific knowledge.

Table 3 shows the characteristic features from the trials involving 20 Swiss teachers as described in the results part of this explorative study. It appears that at primary school level in Switzerland, rather complex FA activities, such as combining several types of student data or several assessment methods in one single inquiry unit, are possible. At upper secondary school level, on the other hand, realistic FA practices could focus on efficiency with simple procedures and a close relation to summative assessment practices.

Table 3. Characteristic features of the trials in the study.

Formative assessment practices at primary school	↔	formative assessment practices at upper secondary school
Several assessment method-competence combinations in one trial	↔	One assessment method-competence combination in one trial
Formative assessment based on various types of student data with observational data being the most common	↔	Formative assessment based on one source of data: written student data
Formative assessment focussed on conducting inquiries	↔	Formative assessment focussed on documenting inquiries
Transversal competences play an important role	↔	Transversal competences play a minor role
Feedback is used to revise the original piece of work	↔	Feedback is transferred to subsequent tasks

For science education in Switzerland, the implementation of FA in IBSE remains a goal to be achieved. The results of the explorative study imply that FA practices which could be realistically expected from a considerable portion of teachers differ between different school levels. This should have consequences for support measures, for example in schoolbooks. Scaffolds in schoolbooks on how to formatively assess IBSE formatively should take into account that the teachers at primary school appear to appreciate FA based on more than one type of student data that also includes transversal competences. But primary school teachers seem to have difficulties in finding indicators for the more abstract competences which cannot be directly observed in the students' hands-on activities. Such specific indicators should therefore be provided in primary schoolbooks. At upper secondary level, teachers seem to value swift FA but seem to have difficulties in assessing student competences that are not based on written data. Specific examples in that respect in schoolbooks might be of assistance.

For the broader education academy, a number of further research questions on implementation processes within the educational system appear to be worth further investigation: The data from the explorative project which lasted for three semesters provided first insights that the quality of the teachers' reasoning on their planning processes, as well as the quality of the trials, developed over the course of the project. In the later interviews, teachers tended to justify their decisions in the planning process with a clearer relation to IBSE theory. Furthermore, the rate of trials which did not fulfil the needs of the project (FA in the context of IBSE) was higher at the beginning and lower at the end of the project. Subsequent projects could investigate the changes in the teachers' assessment literacy in self-directed settings for professional development in more detail, since they appear crucial for the discussion on how innovation could be implemented in the educational system.

Furthermore, there were hints that there might be different implementer types within the teacher community: Some of the teachers tried to reproduce procedures which had once been working, whereas other teachers tried out completely new procedures with every trial. The characteristics of these different implementer types as well as the specific measures of support for further developing these implementer types' teaching deserve being investigated in more detail since they might also have implications for teacher professional development.

Disclosure statement

No potential conflict of interest was reported by the authors.

Funding

This work was supported by European commission (Seventh Framework Programme for Research): [Grant Number 321428].

ORCID

Regula Grob http://orcid.org/0000-0001-6182-8061
Monika Holmeier http://orcid.org/0000-0002-9560-1401

References

Abd El Khalick, F., Boujaoude, S., Duschl, R. A., Lederman, N. G., Mamlok-Naaman, R., Hofstein, A., … Tuan, H. (2004). Inquiry in science education: International perspectives. *Science Education*, *88*(3), 397–419.

Andrade, H. (2010). Students as the definitive source of formative assessment. In H. Andrade & G. J. Cizek (Eds.), *Handbook of formative assessment* (pp. 90–105). New York: Routhledge.

Andrade, H., & Valtcheva, A. (2009). Promoting learning and achievement through self-assessment. *Theory Into Practice*, *48*(1), 12–19.

ARG (Assessment Reform Group). (2002). Assessment for learning: 10 Principles. London: ARG. Retrieved from http://www.assessment-reform-group.org

Artigue, M., & Baptist, P. (2012). *Inquiry in mathematics education. Background resources for implementing inquiry in science and mathematics at school.* Paris: Université Paris Diderot.

Bell, B., & Cowie, B. (2001). The characteristics of formative assessment in science education. *Science Education*, *85*(5), 536–553.

Bell, T., Urhahn, D., Schanze, S., & Ploetzner, R. (2010). Collaborative inquiry learning: Models, tools, and challenges. *International Journal of Science Education*, *32*(3), 349–377.

Black, P. (1993). Formative and summative assessments by teachers. *Studies in Science Education*, *21*, 49–97.

Black, P., & Harrison, C. (2004). *Science inside the black box.* London: GL Assessment.

Black, P., Harrison, C., Lee, C., Marshall, B., & Wiliam, D. (2004). Working inside the black box: Assessment for learning in the classroom. Phi Delta Kappan.

Black, P., & Wiliam, D. (1998). Assessment and classroom learning. Assessment in education: Principles. *Policy and Practice*, *5*(1), 7–73.

Blanchard, M. R., Osborne, J. W., Wallworks, C., & Harris, E. S. (2013). Progress on implementing inquiry in North Carolina: Nearly 1,000 elementary, middle and high school science teachers Weigh In. *Science Educator*, *22*(1), 37–47.

Börlin, J. (2012). *Das Experiment als Lerngelegenheit. Vom interkulturellen Vergleich des Physikunterrichts zu Merkmalen seiner Qualität* [The experiment as an opportunity to learn. From the intercultural comparison of physics education to characteristics of its quality]. Berlin: Logos Verlag.

Börlin, J. & Labudde, P. (2014). Practical work in physics instruction: An opportunity to learn? In H. E. Fischer, P. Labudde, K. Neumann, & J. Viiri (Eds.), *Quality of instruction in physics* (pp. 111–128). Münster: Waxmann.

Burke, K. (2006). *From standards to rubrics in 6 steps.* Heatherton, Victoria: Hawker Brownlow Education.

Bybee, R. (1997). *Achieving scientific literacy: From purposes to practices.* Portsmouth: Heilmann.

Cowie, B., & Bell, B. (1999). A model for formative assessment. *Assessment in Education*, *6*(1), 101–116.

D-EDK Deutschschweizer Erziehungsdirektoren-Konferenz. (2014). *Lehrplan 21* [Curriculum 21]. Luzern: D-EDK.

Dolin, J. (2012). *Assess inquiry in science, technology and mathematics education: ASSIST-ME proposal.* Copenhagen: University of Copenhagen.

EDK Schweizerische Konferenz der kantonalen Erziehungsdirektoren. (1994). *Rahmenlehrplan für die Maturitätsschulen* [Curricular framework for upper secondary schools]. Bern: Schweizerische Konferenz der kantonalen Erziehungsdirektoren.

Euler, M. (2011). *PRIMAS survey report on inquiry-based learning and teaching in Europe.* Kiel: IPN Kiel.

Falchikov, M. (1991). Group process analysis. In S. Brown & P. Dove (Eds.), *Self- and peer-assessment* (pp. 15–27). Birmingham: Standing conference on educational development, paper 63.

Furtak, E. M., & Ruiz-Primo, M. A. (2008). Making students' thinking explicit in writing and discussion. An analysis of formative assessment prompts. *Science Education*, *92*(5), 799–824.

Furtak, E. M., Seidel, T., Iverson, H., & Briggs, D. C. (2012). Experimental and quasi-experimental studies of inquiry-based science teaching: A meta-analysis. *Review of Educational Research*, *82* (3), 300–329.

Greene, J. C., Caracelli, V. J., & Graham, W. F. (1989). Toward a conceptual framework for mixed-method evaluation designs. *Educational Evaluation and Policy Analysis, 11*(3), 255–27.

Grob, Regula. (2017). *Towards the implementation of formal formative assessment in inquiry-based science education in Switzerland. PhD thesis.* Basel: Universität Basel.

Grob, U., & Maag Merki, K. (2001). *Überfachliche Kompetenzen. Theoretische Grundlegung und empirische Erprobung eines Indikatorensystems* [Cross-curricular competences. Theoretical background and empirical validation of a system of indicators]. Bern: Peter Lang.

HarmoS: Konsortium HarmoS Naturwissenschaften. (2008). *HarmoS Naturwissenschaften+. Kompe-tenzmodell und Vorschläge für Bildungsstandards. Wissenschaftlicher Schlussbericht* [HarmoS science education+. Model of competences and suggestions for educational standards. Final scientific report]. Bern: Konsortium HarmoS Naturwissenschaften.

Hartig, J., Klieme, E., & Leutner, D. (2008). *Assessment of competencies in educational contexts.* Göttingen: Hogrefe Publishing GmbH.

Hattie, J. (2009). *Visible learning. A synthesis of over 800 meta-analyses relating to achievement.* London & New York: Routledge.

Hattie, J., & Timperley, H. (2007). The power of feedback. *Review of Educational Research, 77*(1), 81–112.

Heritage, M. (2010). *Formative assessment: Making it happen in the classroom.* Thousand Oaks, California: Corwin Press.

Herman, J. L., Osmundson, E., & Silver, D. (2010). *Capturing quality in formative assessment practice: Measurement challenges, CRESST report 770.* Los Angeles: National Center for Research on Evaluation, Standards, and Student Testing (CRESST).

Husfeld, V. (2009). Aus der Praxis der Leistungsbeurteilung [On the practice of summative assessment]. In D. Fischer, A. Strittmatter, & U. Vögeli-Mantovani (Eds.), *Noten, was denn sonst? Leistungsbeurteilung und -bewertung* (pp. 33–40). Zürich: Verlag LCH.

Jundt, W. (2013). Unpassendes zur Beurteilung [Inappropriate things on assessment]. *profiL Magazin für das Lehren und Lernen, 1*, 10–11.

Kessler, J. H., & Galvan, P. M. (2007). *Inquiry in action: Investigating matter through inquiry.* Washington, DC: American Chemical Society.

Kronig, W. (2009). Schulnoten - Glasperlen des Bildungssystems [Grades – Glass beads of the educational system]. In D. Fischer, A. Strittmatter, & U. Vögeli-Mantovani (Eds.), *Noten, was denn sonst? Leistungsbeurteilung und -bewertung* (pp. 27–32). Zürich: Verlag LCH.

Landis, J. R., & Koch, G. G. (1977). The measurement of observer agreement for categorical data. *Biometrics, 33*, 159–174.

Leahy, S., Lyon, C., Thompson, M., & Wiliam, D. (2005). Classroom assessment: Minute by minute, day by day. *Assessment to Promote Learning, 63*(3), 19–24.

Labudde, P. (2000). *Konstruktivismus im Physikunterricht der Sekundarstufe II [Constructivism in upper secondary physics education].* Bern / Stuttgart / Wien: Haupt Verlag.

Labudde, P. (2007). How to develop, implement and assess standards in science education? 12 challenges from a Swiss perspective. In D. Washington, P. Nentwig, & S. Schanze (Eds.), *Making it comparable: Standards in science education* (pp. 277–301). Münster: Waxmann.

Labudde, P., Nidegger, Ch., Adamina, M., & Gingins, F. (2007). The development, validation, and implementation of standards in science education: Chances and difficulties in the Swiss project HarmoS. In D. Washington, P. Nentwig, & S. Schanze (Eds.), *Making it comparable: Standards in science education* (pp. 235–259). Münster: Waxmann.

Mayring, P. (2010). *Qualitative Inhaltsanalyse. Grundlagen und Techniken.* [Qualitative content analysis. Theoretical background and techniques]. Weinheim: Beltz.

McLoughlin, E., Finlayson, O., & van Kampen, P. (2012). *SAILS – report on mapping the development of key skills and competencies onto skills developed in IBSE: WP 1 – Deliverable 1.1.* Dublin: Dublin City University.

Moskal, B. M. (2003). Recommendations for developing classroom performance assessments and scoring rubrics. Practical assessment. *Research & Evaluation, 8*, 14.

OECD (Organisation for Economic Co-operation and Development). (2005). *The definition and selection of key competences. Executive summary.* Paris, France: OECD Publishing.

OECD (Organisation for Economic Co-operation and Development). (2013). *Synergies for better learning: An international perspective on evaluation and assessment. OECD reviews of evaluation and assessment in education*. Paris: OECD Publishing.

Paris, S. G., & Paris, A. H. (2001). Classroom applications of research on self-regulated learning. *Educational Psychologist, 36*, 89–101.

Priemer, B. (2011). Was ist das Offene beim offenen Experimentieren? [What is open in open experimentation?]. *Zeitschrift für Didaktik der Naturwissenschaften, 17*, 315–337.

Rönnebeck, S., Bernholt, S., & Ropohl, M. (2016). Searching for a common ground – A literature review of empirical research on scientific inquiry activities. *Studies in Science Education, 52*(2), 161–197.

Ruiz-Primo, M. A., Furtak, E. M., Ayala, C., Yin, Y., & Shavelson, R. J. (2010). Formative assessment, motivation, and science learning. In H. Andrade & G. J. Cizek (Eds.), *Handbook of formative assessment* (pp. 139–158). New York: Routhledge.

Schoonenboom, J., & Johnson, R. B. (2017). How to construct a mixed methods research design. *KZfSS Kölner Zeitschrift für Soziologie und Sozialpsychologie, 69*(Supplement 2), 107–131.

Shavelson, R. J., Young, D. B., Ayala, C. C., Brandon, P. R., Furtak, E. M., Ruiz-Primo, M. A., … Yin, Y. (2008). On the impact of curriculum-embedded formative assessment on learning: A collaboration between curriculum and assessment developers. *Applied Measurement in Education, 21* (4), 295–314.

Smit, R. (2009). *Die formative Beurteilung und ihr Nutzen für die Entwicklung von Lernkompetenz. Eine empirische Studie in der Sekundarstufe 1* [Formative assessment and its use for the development of learning competence. An empirical study at lower secondary school level]. Baltmannsweiler: Schneider Verlag Hohengehren GmbH.

Smit, R., & Birri, T. (2014). Assuring the quality of standards-oriented classroom assessment with rubrics for complex competencies. *Studies in Educational Evaluation, 43*, 5–13.

Smit, R., Weitzel , H., Blank, R., Rietz, F., Tardent, J., & Robin, N. (2017). Interplay of secondary pre-service teacher content knowledge (CK), pedagogical content knowledge (PCK) and attitudes regarding scientific inquiry teaching within teacher training. *Research in Science & Technological Education, 35*(4), 477–499.

Stiggins, R. J., Griswold, M. M., & Wikelund, K. R. (1989). Measuring thinking skills through classroom assessment. *Journal of Educational Measurement, 26*, 233–246.

Topping, K. J. (2003). Self and peer assessment in school and university: Reliability, validity and utility. In M. Segers, F. Dochy, & E. Cascaller (Eds.), *Optimising new modes of assessment: In search of qualities and standards* (pp. 55–87). The Netherlands: Kluwer Academic Publishers.

Tsivitanidou, O., Zacharia, Z. C., & Hovardas, A. (2011). High school students' unmediated potential to assess peers: Unstructured and reciprocal peer assessment of web-portfolios in a science course. *Learning and Instruction, 21*, 506–519.

Vögeli - Mantovani, U. (1999). *SKBF Trendbericht Nr. 3: Mehr fördern, weniger auslesen. Zur Entwicklung der schulischen Beurteilung in der Schweiz* [SKBF trend report no. 3: More support, less selection. On the development of educational assessment in Switzerland]. Aarau: Schweizerische Koordinationsstelle für Bildungsforschung.

White, B. Y., & Frederiksen, J. R. (1998). Inquiry, modelling, and metacognition: Making science accessible to all students. *Cognition and Instruction, 16*(1), 3–118.

Widmer Märki, I. (2011). Fächerübergreifender naturwissenschaftlicher Unterricht: Umsetzung und Beurteilung von Schülerleistungen im Gymnasium [Interdisciplinary science education: Implementation and assessment of student achievement at the gymnasium]. PhD Thesis. Basel: University of Basel.

Zachos, P., Hick, T. L., Doane, W. E. J., & Sargent, C. (2000). Setting theoretical and empirical foundations for assessing scientific inquiry and discovery in educational programs. *Journal of Research in Science Teaching, 37*(9), 938–962.

Development of teachers' perspective on formative peer assessment

Iva Zlabkova, Jan Petr, Iva Stuchlikova, Lukas Rokos and Alena Hospesova

ABSTRACT

Teachers' understanding of formative assessment is quite essential for its implementation in everyday classroom practices [Heritage, M. (2007). Formative assessment: What do teachers need to know and do? *Phi Delta Kappan*, 89(2), 140–145]. This study investigates the development of teachers' perspective on formative peer assessment when it is newly introduced into their teaching. A group of 18 science teachers participated in design-based research project in which they developed and enacted inquiry-based teaching units with embedded formative peer assessment instances. Teachers' perspective on formative peer assessment was captured by means of group discussions and reporting tool at different stages of the project (before teaching the experimental units, after having taught the units and 20 months after that). Data were analysed by thematic analysis. The findings show that immersing teachers into this relatively new method, formative peer assessment, especially when they are co-investigating their own practice (in research project), may have significant impact on their conceptualisation and implementation of formative assessment in general.

Introduction

Better learning of science can be achieved if science is taught in an investigative approach (Hazelkorn et al., 2015; Rocard et al., 2007) and if students' learning is facilitated through formative assessment (Black & Wiliam, 1998; Heritage, 2007; Shepard, 2005).

Though the investigative approach is understood as a key instructional strategy in science (e.g. Abell & Lederman, 2007), its adoption in everyday science teaching is not always easy, even if teachers are willing to employ the inquiry approach. One of the most salient problems that teachers see in inquiry-based science teaching is the proper assessment of the students' learning process and learning outcomes (Dolin & Evans, 2013; Stuchlíková, Petr, & Papáček, 2013; Wilcox, Kruse, & Clough, 2015). Inquiry-based teaching aims to develop students' competences of questioning about the natural world, generating hypotheses, designing investigations, collecting data and analysing them in order to resolve the question (Windschitl, 2003). This is not a linear or straightforward process, and it is, therefore, essential to assess students' progress formatively, concerning their current state, already mastered skills, knowledge and understanding, as well as the steps needed going forward.

To use formative assessment successfully in the classroom, teachers 'must view formative assessment as a worthwhile process that yields valuable and actionable information about students' learning. … must view formative assessment and the teaching process as inseparable and must recognise that one cannot happen without the other … ' (Heritage, 2007, p. 145). Though this can be agreed, the teachers' perspective on formative assessment should be investigated in more depth, especially as it develops when teachers try to implement complex inquiry-based tasks in their teaching.

As a part of the research programme ASSIST-ME, in which formative assessment in inquiry-based science teaching was studied, this study focused on teachers' experience with formative peer assessment. This assessment method was new for participating teachers and their students, as peer assessment is not common in Czech teaching culture (Šeďová, Švaříček, & Šalamounová, 2012).

The main aim of this investigation was to explore the impact that teachers' participation in research and experimental teaching, where formative peer assessment was embedded, may have on teachers' perspective on formative peer assessment, and how this perspective further develops when the teachers continue to use formative peer assessment in their practice.

Theoretical background

Changes of teaching practice and teachers' perspective on teaching

The implementation of a new approach and changes in teaching practice are challenging tasks, as the teachers tend to keep practices which function well for their students and are reluctant to introduce new ones unless they are persuaded about their higher effectiveness (Birman, Desimone, Porter, & Garet, 2000; Borko, 2004; Guskey, 2003; Guskey & Yoon, 2009).

Professional development programmes should, therefore, strive to provide experience and evidence that can influence teachers' conceptualisation of their teaching (Fullan & Hargreaves, 1992; Wagner, 2003). It is crucial to provide the teachers with enough evidence and good practice of a newly introduced approach (Anderson, 2002; Sunal et al., 2001; Windschitl, 2003). Participation in research on such a new approach can provide such evidence and good practice.

Within the domain of science education, the model of research experiences for teachers (e.g. Blanchard & Sampson, 2018; Cooper & Cowie, 2010) is used to provide the teachers with good, personally relevant practice in science, including science pedagogy. This research experience shows the positive impact on the teachers' beliefs about teaching, on the amount of pedagogical discontentment they felt, on their self-efficacy, and for those who participated in the science pedagogy-focused programme, also the impact on changing the design of inquiry-based lessons.

Formative assessment

The essence of formative assessment lies in the teachers' ability to utilise the 'moments of contingency'[1] (Black & Wiliam, 2009) for facilitating students' learning, providing them with proper feedback and with suggestions for future learning at the right time. Though the teachers, in general, understand and value these instructional components, they, of course also see the limits, challenges and obstacles[2] in enacting them (Dolin & Evans, 2018; Evans et al., 2018; Kingston & Nash, 2011).

Black and Wiliam (1998) described formative assessment as teachers' and students' activities which provide feedback information that is used to modify teaching and learning. In their paper on the theory of formative assessment (FA), Black and Wiliam (2009) stated that 'formative assessment can be conceptualised as consisting of five key strategies: (1) clarifying and sharing learning intentions and criteria for success; (2) engineering effective classroom discussions and other learning tasks that elicit evidence of student understanding; (3) providing feedback that moves learners forward; (4) activating students as instructional resources for one another; and (5) activating students as the owners of their own learning' (p. 4).

Formative peer assessment

Peer assessment is an arrangement in which peers of similar status consider the quality of each other's learning outcomes or products (Topping, 1998). Peer assessment is an appraisal of not only the outcomes of learning but also of the processes of learning (Van Gennip, Segers, & Tillema, 2009). When peer assessment is implemented as a tool to support learning, it is an integrated part of a collaborative learning process (Van den Bossche, Gijselaers, Segers, & Kirschner, 2006) and can function as a formative pedagogical tool (Topping, 2009). Formative peer assessment stimulates the learning of higher-order skills such as sharing responsibility, reflection, discussion, and collaboration (Heitink, Van der Kleij, Veldkamp, Schildkamp, & Kippers, 2016; Strijbos & Sluijsmans, 2010; Van Zundert, Sluijsmans, & Van Merriënboer, 2010). According to Strijbos and Wichmann (2018), formative peer assessment has become a popular instructional approach during the past two decades. Still, it is not yet broadly used in every teaching culture.

Implementation of formative peer assessment is related to the student-centred teaching practice. If students are going to be successfully involved in monitoring and assessing their own and their peers' learning, then they need to be regarded by their teachers as partners in learning (Heritage, 2007). The potential of formative peer assessment for establishing a formative culture of learning in the classroom (Shepard, 2000) is therefore probably stronger than an impact of formative feedback provided only by the teacher.

Research questions

More than just a method of assessment, formative assessment is rather an approach to teaching, which modifies teaching practice as well as students' learning. For teachers, it means to monitor students' learning processes continually and recurrently adapt their teaching actions to them. For students, it means to become more active, both as an owner of one's learning and as a peer-learner, i.e. member of the learning community in the classroom. To facilitate these formative processes, teachers need some impetus and/or support to change their attitude towards assessment and to the facilitation of students' responsibility for learning.

The design-based ASSIST-ME research project[3] provided such an impetus as teachers interested in up-scaling formative assessment worked in teams with researchers to design, implement and evaluate assessment methods in three rounds of experimental teaching and research distributed over a year and a half. The main objective of this study was to examine teachers' views on formative peer assessment after experimental implementation and research on it and after a more extended period of practising it. Therefore, the research questions guiding this study were:

How did the teachers participating in research on formative assessment of inquiry-based science learning perceive formative peer assessment after they implemented formative peer assessment instances in experimental teaching units; what they saw as strengths and weaknesses, opportunities and challenges, potentials and obstacles?

How the teachers' perception of formative peer assessment in inquiry settings has developed in a more extended period after their participation in research, supposing they continued using formative peer assessment in their teaching?

Method

Participants

In total, 18 volunteering teachers were involved in our study, participating as members of a Czech research team in European project ASSIST-ME. Teachers' role was to develop, together with researchers, experimental teaching units with embedded formative peer assessment and enact them, reflect the process of implementation, co-operate with teacher colleagues on the refinement of teaching units and provide information about their experience from experimental teaching practice.

Divided into three smaller cohorts (6 biology teachers at primary level, 6 biology teachers at lower-secondary level, and 6 mathematics teachers at primary level), all of them participated in three subsequent rounds of experimental teaching (once as acting teacher, twice as co-teacher developing the unit). The whole group worked together at the beginning of the project, elaborating the concept of formative assessment and formative peer assessment and regularly met after each round of experimentation and research to reflect their teaching and peer assessment related experience. This structure of the project provided opportunity for all participating teachers to make comparable experience with introduction of formative peer assessment into their teaching.

The content criteria for creating teaching units were twofold; the inquiry-based tasks for students should be accompanied by formative peer assessment. Teachers were encouraged to suggest the topic and proper tasks. Based on it, the teaching units were elaborated in co-operation with researchers so that the inquiry tasks were of comparable type of inquiry and criteria for the assessment of peers' products were clear. Even if co-operating intensely, the teachers kept full ownership of their instructional decisions related to particular realisation of formative peer assessment in their classrooms. The quality of implementation and the quality of formative peer assessment itself was discussed with participating teachers immediately after their teaching, enabling them to reflect in-depth currently experienced strengths, weaknesses, challenges and limits of this assessment method.

Procedure and data collection

At the beginning of the project, teachers' previous experience with formative assessment was revealed through a structured group discussion method.

The teachers then developed with researchers experimental teaching units with embedded instances of formative peer assessment (FPA) of different kinds (see topics and forms of FPA in Table 1). Experimental teaching units consisted of one to six inquiry tasks of different complexity, which usually took two months of teaching.[4]

All 18 participating teachers gained varied experience with formative peer assessment, both as an acting teacher and as a co-teacher developing the units and discussing the experience of other teachers. At the end of the experimentation period, all teachers answered the reporting tool collecting the data on teachers' perspective on formative peer assessment. As the group is relatively small and subdivided into cohorts, and the perspective on formative assessment is subjective and contextualised, the data are treated as qualitative, though some particular quantification on closed and scaled items is utilised.

Participating teachers were contacted again after twenty months of their independent practice, and asked to participate in follow-up interviewing. Eight of them (four from lower-secondary level, four from primary level – two from biology, two from the mathematics group; see Table 1) who agreed to answer a follow-up questionnaire created a follow-up group. This follow-up group was balanced with respect to subjects and school level, and its representativeness for the original group was explicitly assessed. The resting teachers either were not working in the same school (2 of them were on maternity leave, 3 has changed the school) or not respond to our call or were hesitating to co-operate (for unspecified reasons, some of them mentioned the lack of time, but the reason could be also limited use of formative assessment in their practice).

For the sake of comparison of teachers' views on FPA in Time 1 (after completion of experimental teaching period) and Time 2 (twenty months later) the same reporting tool was administered to follow-up group (except the last question, which was related directly to the experimental teaching). The overall view of the collected data and the methods used is presented in Table 2.

Measure

Teachers' Reporting Tool (TRT) was developed in the ASSIST-ME project and was used for collecting the teachers' opinion on the experimental implementation of particular formative assessment methods. As such, the tool was of qualitative nature, though it used some quantifications (ratings). It's first part consisted of eleven Likert scale statements

Table 1. Experimental teaching.

Group/Subject	Number of teachers	Educational level	Peer assessment format	Taught topic	Number of taught classes	Follow-up interviews
Primary mathematics	6	Grades 2, 4, 5	Group discussion (2nd grade), pairs or small groups	Basic geometrical shapes and their area, big numbers	6	2
Primary biology	6	Grades 3, 4	Individual	Germination	6	2
Secondary biology	6	Grades 6, 7, 8, 9	Individual	Germination Human physiology	6	4

Table 2. Data collection methods.

	Before experimental teaching	After the experimental teaching	Follow-up
Methods	Group discussion	Structured questionnaire (ASSIST-ME reporting tool)	Structured questionnaire (ASSIST-ME reportig tool)
Number of participants	18	18	8
Aim	Revealing experience in and attitudes to FA in inquiry-based teaching	Perspective on FA after experimental teaching, impact of research experience	Reflection of the role of FA in one's teaching, changes in perspective on FPA in the long term

related to the enactment of particular formative assessment method and in the second open-ended questions part, the respondents were asked to share their opinion on strengths and weaknesses, opportunities and challenges of the assessment method and its potentials and obstacles related to its implementation (see Figure 1 for statements). According to our knowledge, any other tool was published which would be similar enough for establishing concurrent validity of TRT. The face validity of the TRT is nonetheless relatively high as it directly asks what is intended to be measured, i.e. teachers' perspective on the implementation of a formative assessment method. Content validity of the Teachers' Reporting Tool (the term questionnaire will be used for the brevity from here onwards) is supported by addressing different aspects of teachers' view on the implementation, i.e. realisation (time, easiness of implementation), relation to current teaching practice, purpose of formative assessment, and the students' behaviour.

Data analysis

The group discussion was recorded and all mentioned themes and meanings were listed in minutes, which were approved by teachers and then summarised as a thematic report about group discussion.

Data from the Time 1 questionnaire were analysed in two stages. The ratings of statements were treated as ordinal data, and relative proportions of different levels of agreement within the whole group of teachers were calculated for particular statements. The second part of the questionnaire elicited open-format answers which were treated as qualitative data, transcribed and open-coded. The codes were then sorted into emerging thematic categories, based on semantic closeness, thus also allowing for single-code categories, which ensured that no idiographically important meanings remained hidden.

The analyses of data from Time 2 aimed to reveal changes in development from Time 1 to Time 2. The ratings from Time 2 were therefore treated as repeated measurements; their prospective change was analysed quantitatively (by a marginal homogeneity test) and then in an idiographic manner concerning changes in particular statements and for particular teachers. The answers to open questions in the second part of the questionnaire were also analysed from the perspective of a change in time. The answers from Time 2 were open-coded, and codes were sorted into emerging thematic categories separately for Time 2. Then the Time 1 and Time 2 categories were compared. As the group of teachers was very small, the overall number of coded quotations was too small for any quantitative comparison and therefore visual comparison (Reavey, 2018), networks of thematic categories, were used for particular questions.

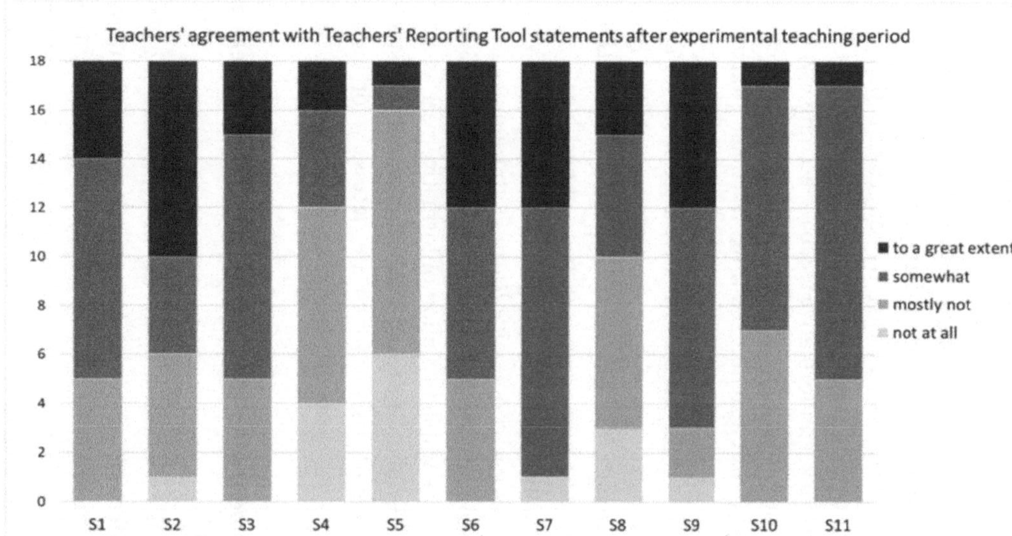

Legend: Number of teachers (vertical axis) who agree with the statements:

Formative peer assessment

S1: was easy to implement.
S2: took more time than I expected.
S3: is something I could easily employ in my teaching approach.
S4: is something I do employ in my daily teaching practice.
S5: is consistent with the typical teaching practice in my country.
S6: could potentially serve summative assessment purposes.
S7: could produce information that may inform my instructional decisions about my next steps in the unit.

The students

S8: were used to the peer assessment method.
S9: were keen to engage in the peer assessment method.
S10: were able to reflect on the merits of the artefact they were assessing.
S11: were able to improve their artefact as a result of the feedback comments.

Figure 1. Teachers' ratings of their experience with formative peer assessment.

Results

Before the experimental teaching – teachers' preceding experience in formative assessment

At the beginning of their participation in the research project, the teachers watched an attuning video of formative assessment in a physics inquiry-based lesson[5] and then discussed in a group the following topics: the use of formative assessment methods in their inquiry teaching, the possibility of using the information gathered in formative assessment for summative purposes, and the challenges that they saw in using formative assessment in inquiry teaching.

Some of the teachers had difficulties with the conceptualisation of formative assessment as the term itself was relatively new to them (usually immediate feedback or feedback for learning are the terms used by Czech teachers). During the discussion, the teachers described instances of formative assessment in their own practice, mostly on-the-fly formative feedback, written formative feedback, and formative feedback accompanying summative scores on a test. All teachers admitted that they prefer summative assessment, because it is strongly rooted in the Czech teaching culture; teachers, students and parents are used to and expect to receive the marks.

The teachers did not see the greater possibility to use formative assessment for summative purposes and saw the problem in transparency of the overall assessment (Teacher 5: '*A teacher should declare to students in advance how the formative assessment results will be used in summative scores. And it is not easy and sometimes even not possible*').

Teachers admitted that peer assessment is relatively new for them, though they valued peer communication as an important part of cooperative learning and inquiry learning. The teachers said that as far as the students are not used to being in an assessor role, they had foreseen challenges related to the implementation of formative peer assessment in their teaching. They worried especially about time demands, the passivity of students, the lack of some students' knowledge and/or skills which are necessary for inquiring, investigation and assessment.

Teachers' view on formative assessment after their experimental employment of formative peer assessment methods

All the teachers answered a reporting tool after finishing the experimental teaching period. The questionnaire had two sets of statements. The teachers rated their agreement on the four-point Likert scale from 'not at all' to 'to a great extent'. The first set of statements deals with the employment of FPA and its value for summative assessment purposes and for the directing of further steps in teaching. The second set relates to the teacher's view on students' engagement and on the improvement of their learning. Answers to this rating part of the questionnaire are summarised in Figure 1.

Figure 1 shows that when rating FPA, the teachers agreed the least with the statements reflecting the common use of FPA in their teaching, (statements 4 and 5). On the contrary (in statements 1, 3, 6 and 7), teachers mostly agreed (at least to some extent) that FPA was easy to implement (13 teachers), could be easily employed in their approach to teaching (13 teachers) and could both, produce information for summative assessment of students' learning (13 teachers) and for their instructional decisions (17 teachers).

When teachers evaluated the impact that FPA had on their students, they reflected that the students are not used to peer assessment methods (10 teachers in statement 8), but agreed (in statements 9 and 11) that the students were keen to engage in peer assessment (15 teachers) and were able to improve their learning as a result of peer feedback (13 teachers).

The answers to the open questions in the second part of the questionnaire were coded by the third and fourth author (the inter-coder reliability was calculated as a agreement of 82% (Popping, 1988) and the discrepancies were discussed until reaching congruent coding), the resulting categories are summarised in Table 3.

The teachers reflected more strengths than weaknesses of peer feedback as a method of formative assessment. Among the strengths, they especially saw the potential to deepen the learning of students, facilitating their mastery of critical thinking and development of their cognitive abilities. The weaknesses were considered to be higher time demands and the relative unpreparedness of students for being assessment agents, e. g. students' problems to properly formulate the feedback and the risk of projection of bad personal relations into the provided feedback.

When teachers commented on the opportunities and challenges that they identified in the employment of this assessment method, the most represented category of opportunities was the broad scope of the prospective usage of FPA, also in other teaching contexts than inquiry and other subjects that they are teaching (e.g. physical training, art education). The challenges are associated with arrangements, which could at least partially alleviate the time demands related to the time-consuming nature of peer feedback, and with the need to develop FPA skills since the early years of schooling.

The category covering a high number of codes for overcoming the challenges were collegial support (which included also support from researchers) and the frequent use of FPA.

Concerning what the teachers would change in the structures and procedures that they employed, ten different categories emerged from the codes. The most frequent category represented the relative satisfaction of teachers with their particular way of implementing FPA (nothing or almost nothing should be changed); several teachers also shared the opinion that simple tasks and smaller steps should be better.

Further development of teachers' perspective on formative assessment

The second research question aimed at the further development of teachers' perspective on formative assessment, and especially on formative peer assessment, during the follow-up period in which they had been trying to use formative assessment in their science and mathematics lessons. To answer the second research question, follow-up data from teachers who agreed to participate ($N = 8$) were collected 20 months after finishing the experimental teaching period. The follow-up group consisted of four primary and four secondary teachers, and the representativeness of this small follow-up group of teachers for the whole group of eighteen teachers participating in the research was estimated by comparison of the ratings in the first administration of the questionnaire by Fisher's exact test. The Fisher's exact test p-values ranged from 0.10 to 0.99 for particular questions, which means that the ratings of the follow-up group's did not significantly differ from the ratings provided by the rest of the teachers.

Table 3. Teachers' reporting on their experience with formative peer assessment implementation – open questions.

Questions	Categories of codes (number of codes in the category)
1. What are the main strengths of peer feedback as an assessment method?	deeper learning, mastery (11) critical thinking, improvement of cognitive processes (10) development abilities to assess (5) peers as resources (5) improvement of language abilities (more precise description and explanation) (5) students' activity stimulation, motivating (4) development of self-regulation skills (4) changes in teaching (3) development of inquiry skills (2) improvement of social relations in the classroom (2)
2. What are the main weaknesses of peer feedback as an assessment method?	time demands (7) the projection of bad personal relations into the feedback (6) lack of students' vocabulary, problems to formulate feedback (6) exceeding some students' abilities to assess (5) lack of content knowledge (4) worry about quality of peer feedback (4) low engagement of some students (2)
3. What opportunities and challenges do you identify in enacting this assessment method?	training of FPA skills is necessary, start in primary education (8) broad scope of usage (beyond inquiry and in other subjects) (7) need to solve time demands (7) students are not used to working with critical feedback, a good climate is necessary (4) FPA needs a long time be fully accepted as helpful (3) improvement of classroom climate (2) very suitable for inquiry learning assessment (2) use of technologies, more structured FPA (2) lack of materials for students (2) smaller amount of students in the class needed (2) possibility to connect formative and summative assessment (1) students can have problems with the expression of the feedback (1) FPA helps students to avoid or correct their own mistakes (1) need to control correctness of feedback (1)
4. What resources could help teachers overcome these challenges?	collegial support (attendance classes as an auditor, discussions with researchers etc.) (6) frequent use of FA (6) examples of good practice (3) more attention to benefits of FA in curriculum composition (3) implementation of FPA at primary education level (3) cooperation with researchers (preparation of materials, help with implementation) (3) preliminary training of assessing skills including proper vocabulary (2) small amount of students in a class (2) materials for teachers with examples (1)
5. What would you have done differently, in terms of the structure and procedures you have employed, if you had the opportunity to repeat the implementation of peer feedback?	nothing or almost nothing should be changed (6) simpler tasks, smaller steps (4) more time for explanation and training of peer assessment (3)

(*Continued*)

Table 3. Continued.

Questions	Categories of codes (number of codes in the category)
	allow students to discuss more about the assessment (2) more attention to contradictory evidence in experiments (1) start FPA in non-inquiry tasks at first (1) more of a connection between FA and SA (1) actions to prevent incorrect feedback (1) oral feedback before written forms (1) facilitate participation of passive students (1)

The follow-up group teachers described different patterns of formative assessment use in their teaching practice between Time 1 and Time 2 (see Table 4). It is visible from these reports that some teachers kept conceiving FPA methods as pure methods (e.g. Teachers A, B, C); others see FPA as an integral part of their teaching approach (e.g. Teachers G, H).

These teachers answered the same questionnaire once again in Time 2. Data from the first part of this second administration were analysed again in two stages. At first, the ratings of teachers' agreement with eleven statements of the questionnaire were analysed as group data to see whether there occurred any significant changes in teacher' ratings.

As the group was very small ($N = 8$) for a within-subjects analysis which involves repeated measures (i.e. pre–post design), we had to aggregate the data. We, therefore, approximated individual ratings of particular teachers as independent and equally valued measures and aggregated them into one data pool of agreement/disagreement statements (8 teachers by 11 statements, i.e. 88 ratings). We then used the Stuart-Maxwell statistic for testing marginal homogeneity (Sun & Yang, 2018) which refers to equality (lack of significant difference) between the distribution of ratings in Times 1 and 2. If there is a change overall, either a decrease or increase from one time-point to another, relative to what is expected by chance, the result will be significant. The related-samples marginal homogeneity test standardised statistic was $MH = 1{,}964$, $p = .05$, just reaching the conventional level of significance, showing that there is a statistically significant difference in teacher's ratings in Time 1 and Time 2. To see in which aspects the teachers expressed a different opinion, we had to analyse the individual item data. The teachers' ratings in Time 1 and Time 2 and the change of ratings are presented in Table 5.

Table 4. The follow-up group teachers' use of formative assessment in their everyday teaching practice during the follow-up period.

Teacher	Use, FPA and other methods of FA
Teacher A	More on-the-fly FA, FA complement SA (marks) in oral examinations, FPA when evaluating the outputs of small projects or inquiry tasks
Teacher B	Regular use of FA, not only in inquiry lessons (with respect to abilities of students in a particular class – whether they are able to assess formatively)
Teacher C	Oral FPA in physical education lessons, in science FPA used only about twice, self-assessment used regularly, esp. in oral examinations
Teacher D	FPA used in rehearsals at the end of lessons
Teacher E	FPA very successful for younger students in assessment of products in art education, FPA regularly in all subjects when evaluating group work
Teacher F	Self-assessment every month, FA used regularly
Teacher G	FA used continuously in my instruction
Teacher H	FA as a regular part of my teaching

Table 5. Teachers' ratings in Time 1 (T1) and Time 2 (T2) and change (T1–T2).

	Teacher A			Teacher B			Teacher C			Teacher D			Teacher E			Teacher F			Teacher G			Teacher H			Group sum of changes
Statements	T1	T2	Change	T1	T2	Change	T1	T2	Change	T1	T2	Change	T1	T2	Change	T1	T2	Change	T1	T2	Change	T1	T2	Change	
S1	3	3	0	3	2	−1	2	3	1	3	3	0	3	3	0	2	4	2	3	4	1	3	4	1	3
S2	2	3	−1	2	3	−1	1	3	−2	4	3	1	2	3	−1	4	1	3	2	2	0	3	3	0	−1
S3	3	2	−1	3	2	−1	3	4	1	3	3	0	3	4	0	2	1	−1	4	4	0	3	3	0	−2
S4	3	3	0	2	2	0	2	3	1	2	2	0	2	2	0	4	4	0	3	4	1	2	4	2	3
S5	1	1	0	2	2	0	2	2	0	2	2	0	2	4	0	4	3	−1	2	3	1	2	2	0	−1
S6	3	3	0	2	3	1	3	4	1	3	3	0	4	4	0	2	3	1	4	2	−2	4	2	−2	1
S7	4	4	0	3	3	0	4	4	0	3	3	0	4	4	0	1	3	2	3	4	1	4	3	−1	1
S8	1	1	0	2	1	−1	2	3	1	1	2	1	2	3	1	2	4	2	3	3	0	2	3	1	5
S9	2	3	1	3	3	0	4	3	−1	4	4	0	4	3	0	1	3	2	3	3	0	3	3	0	2
S10	2	3	1	2	2	0	3	2	−1	3	3	0	3	2	0	4	3	−1	3	3	0	2	3	1	0
S11	3	2	−1	3	3	0	2	3	1	4	4	0	3	3	0	3	3	0	3	3	0	2	3	1	1
Sum of changes			−1			−3			2			2			0			9			2			3	12

Legend: Rating scale was:1-notatall,2–mostlynot,3–somewhat,4-togreatextent. As the agreement with S2 means a less positive view on FPA, the change scores were reversed, so that all the positive values of change indicate more positive view. Negative values of changes mean a shift towards a more negative view.

Statements:

S1: Formative peer assessment was easy to implement.
S2: Formative peer assessment took more time than I expected.
S3: Formative peer assessment is something I could easily employ in my teaching approach.
S4: Formative peer assessment is something I do employ in my daily teaching practice.
S5: Formative peer assessment is consistent with the typical teaching practice in my country.
S6: Formative peer assessment could potentially serve summative assessment purposes.
S7: Formative peer assessment could produce information that may inform my instructional decisions about my next steps in the unit.
S8: The students were used to the peer assessment method.
S9: The students were keen to engage in the peer assessment method.
S10: The students were able to reflect on the merits of the artefact they were assessing.
S11: The students were able to improve their artefact as a result of the feedback comments.

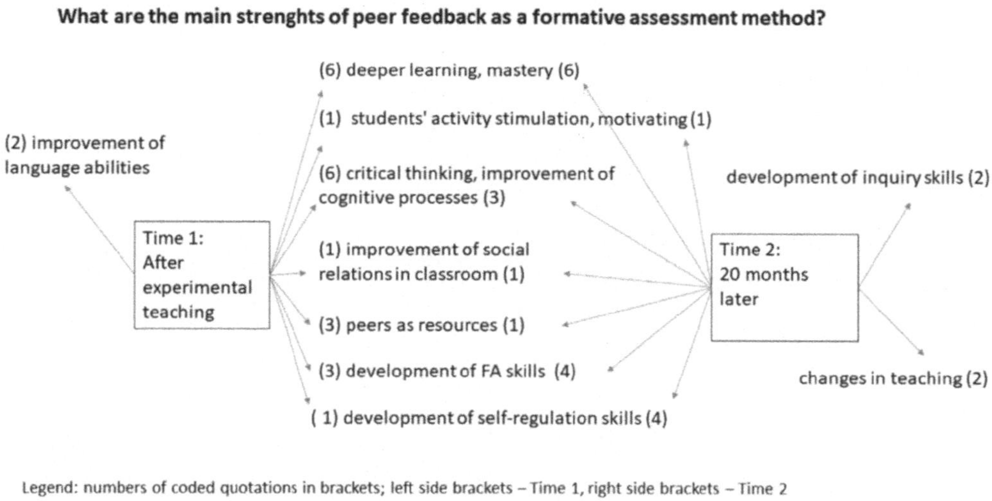

Figure 2. Thematic categories in Time 1 and Time 2 for main strengths of peer feedback as a formative assessment method.

As can be seen from Table 5, the changes in teachers' perspective on FPA were different; some of them mostly did not change their opinion (Teacher E), some became more optimistic (Teacher F) and some of them slightly more sceptical (Teacher B). For questions, the most noticeable change showed the ratings of statement 8; teachers became more positive that their students were used to the peer assessment. This is in correspondence with teachers' report of FPA use in their teaching practice. The statement in which the teachers as a group slightly lowered their agreement concerned easiness to employ FPA in their teaching approach, which corresponds with some other findings in the analysis of open questions (see below – low support from colleagues, the reluctance of parents).

The answers to the open questions of the questionnaire in Time 2 were open-coded and thematically categorised independently of the coding categories of Time 1. Thus, several new thematic categories have emerged. The categories for Time 1 and Time 2 were summarised in networks of categories, which allows visual comparison.

Figure 2 shows the networks of thematic categories for the first open question.

This group of teachers who reflected on FPA immediately after their experimental teaching and then with twenty months of further practice (with various amount and formats of FPA) has in the second time left the category related to the improvement of language abilities. The prevalence of the category 'critical thinking, improvement of cognitive processes' has also dropped down. On the contrary, the category 'development of self-regulation skills' became more salient and two new thematic categories had emerged: 'development of inquiry skills' (example of coded quotations: '*FPA develops students' ability to formulate and defend their own opinions, provide and accept constructive criticism; it leads to greater responsibility for their own work*', Teacher B) and 'changes in teaching' ('[trials to use FPA yielded] *Mainly a change in what I am doing and a different view on formative assessment; it fits into all subjects*', Teacher A).

The comparison of thematic categories of the main weaknesses of FPA, as seen by this group of teachers in times 1 and 2 is presented by Figure 3. New thematic categories which

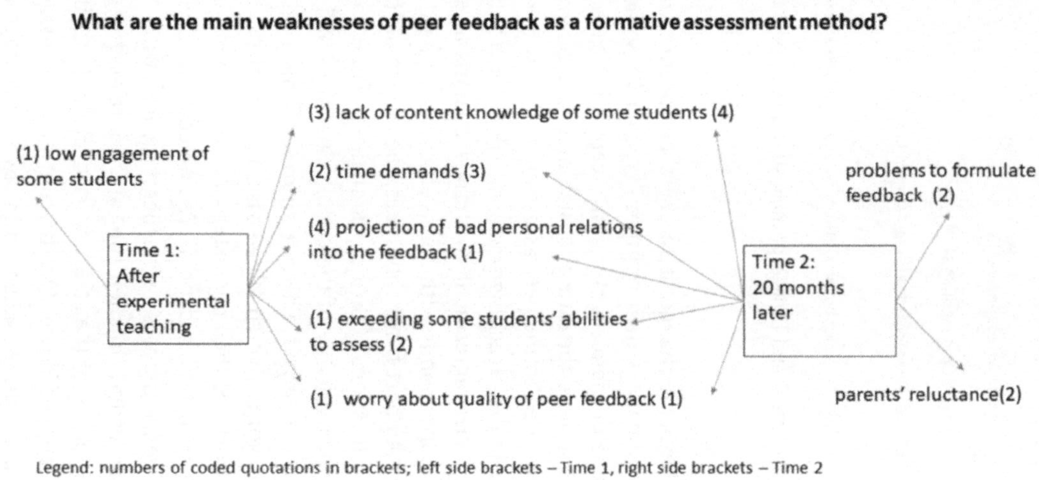

Figure 3. Thematic categories in Time 1 and Time 2 for the main weaknesses of peer feedback as a formative assessment method.

emerged during the time course were 'problems to formulate feedback' ('[they are] *not too good in formulation; for children it is necessary to lead them to precise expressions so that everyone understands what they want to communicate*', Teacher G) and 'parents' reluctance' ('*parents often do not understand the purpose* [of FPA]; *they expect more marks*', Teacher H). Another marked feature is a weakening of the category related to the projection of bad personal relations into the feedback.

A comparison of teachers' views on opportunities and challenges identified in the practice of FPA in Time 1 and Time 2 is summarised in Figure 4. The low understanding of colleagues and challenging preparation for teaching has emerged after more extended practice ('*Colleagues do not understand FA, I feel a need to defend FA in front of them*', Teacher H; '*It is really challenging to prepare, it requires a lot of thinking*', Teacher B)

The last comparison of teachers' views on FPA in Time 1 and Time 2 concerns resources and practices that could help the teachers to overcome the challenges that they identified in the employment of FPA in their teaching. Immediately after the experimental teaching, teachers reported the importance of the frequent use of FPA ('*More frequent use of this method would also lead to some routines in organising of it and thus it would not be so time-consuming for teachers. Everything is about the teacher's will to do the job differently, even if it is time-consuming*', Teacher A).

Cooperation with researchers who had helped them with preparation and enactment of FPA instances was also mentioned as salient ('*Support of researchers was important. It helped to discuss it* [the experimental teaching] *with yet other partners than other teachers*', Teacher G). After further FPA practice on their own the teachers pointed at further different aspects, also at the importance of structure and conditions for assessment (materials, when and how, number of students in the class etc.) (Figure 5).

General discussion

Teachers in this study participated in a research project aiming at the development of formative assessment methods in inquiry-based science teaching, and at the very beginning of the research, they had had limited experience with formative peer assessment. Though some of them have been using some methods of formative assessment in their preceding practice, formative peer assessment was a method which was new for them. The first research question of this study was focused on how these teachers perceive formative peer assessment methods after they, as the members of the research team, have enacted experimental teaching units with embedded formative peer assessment instances; what they saw as strengths and weaknesses, opportunities and challenges, potentials and obstacles. Our findings show that the experimental teaching of approximately two-month length provided teachers with FPA experience that stimulated the expansion of their perception of formative assessment. They started to perceive FPA as something that they can employ in their daily teaching practice and as something that could produce information that may inform their instructional decisions about the next steps in teaching, which is an essential part of formative assessment (Heritage, 2007).

Their experience with FPA was broadly conceptualised, which also mentioned by Andersson and Palm (2017). The teachers considered the cognitive, motivational and emotional, as well as the personal and social context of FPA as can be seen from Table 3. The teachers reflected more strengths than weaknesses of peer feedback as a method

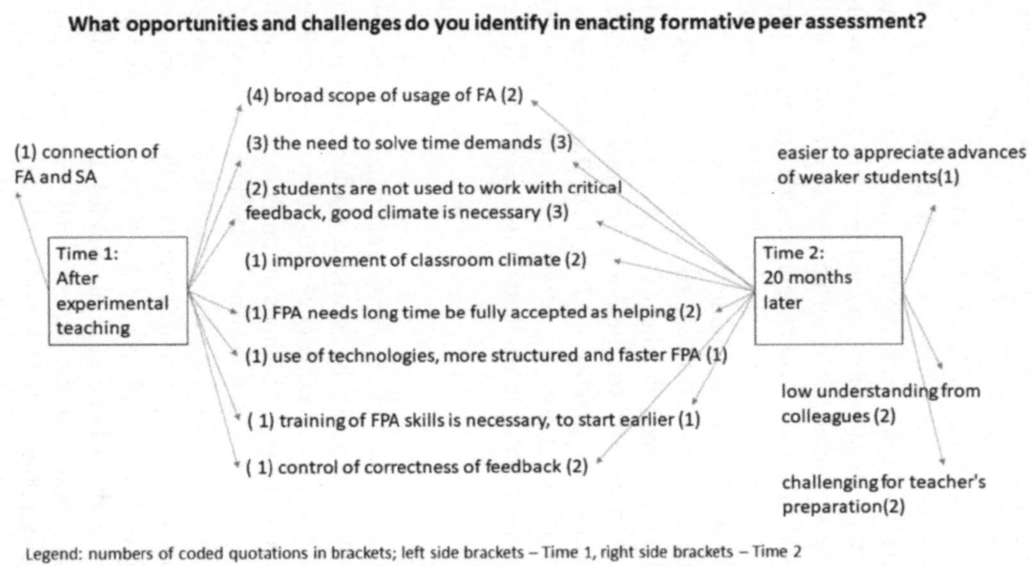

Figure 4. Thematic categories in Time 1 and Time 2 for opportunities and challenges of peer feedback as a formative assessment method.

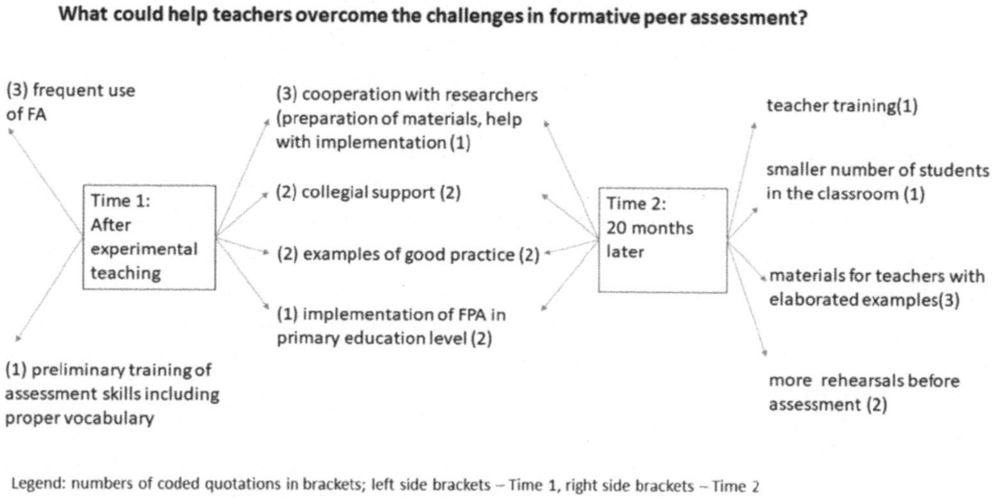

Figure 5. Thematic categories in Time 1 and Time 2 for possibilities to overcome challenges in the implementation of peer feedback as a formative assessment method.

of formative assessment. They appreciated the potential of FPA to facilitate mastery in learning and critical thinking but also reflected the importance of conditions for successful deployment of FPA - a positive social climate in the classroom and time for training of students' skills for assessment.

What teachers nonetheless left unmentioned is the relation of FPA practice to learning goals and the question of clarifying criteria and sharing learning intentions and criteria for success with students (Black & Wiliam, 2009). It turned out to be an essential problem in the studied implementation of FPA, as the teachers in our study repeatedly mentioned that assessment was too difficult for some students, which indicates that teachers were not entirely successful in helping the students to internalise quality criteria (Shepard, 2005).

The teachers valued their research experience. All of them volunteered, so they were keen to learn something new. They mostly mentioned that they would not start trying FPA in their teaching without researchers' support, and on the other hand, they admitted that the use of formative assessment cannot be taught, but can be only constructively learned, which comprises changes in teachers' beliefs. Similar findings were published by Blanchard and Sampson (2018) who stated that research experience is assumed to have a significant impact on teachers' beliefs and instructional decisions.

The second research question focused on the development of the teachers' perception of formative peer assessment in inquiry settings in a longer period after their participation in research, supposing they continued using formative assessment in their teaching. The long-term consequences of the research experience and subsequent practice of formative peer assessment of participating teachers were reflected by the follow-up part of the study. It has shown that teachers from the follow-up group continued to use formative assessment, not only FPA but also other methods. The described changes in teachers' perception of FPA indicated a generally more positive teachers' view on formative assessment after twenty months, though teachers also perceived that formative peer assessment is not easily employable in their teaching approach, especially with little understanding from colleagues, insufficient support, and the reluctance of parents. These findings are in correspondence with a study on the employment of inquiry-based science teaching (Heinz et al., 2018), where a similar impact of culturally situated teaching context is also reported.

These teachers have changed their view on formative assessment and peer assessment depending on their practice of formative assessment and the experience they gained. They reflected different issues, but as a group, they showed a more thematically differentiated perspective on formative peer assessment at the time of the second measurement. Comparison of the data sets collected after experimental teaching and twenty months later proved that teachers started to recognise further aspects of formative assessment in general. These were, among others, the impact of formative assessment on the development of inquiry skills and on self-regulation, which was described in Black's and Wiliam's theorising (Black & Wiliam, 2009). Bryant and Carless (2010) published similar findings. Some of the teachers also realised that the practice of formative peer assessment in their instruction changes their approach to teaching, which was also found in the study of Box, Skoog & Dabbs (2015).

Limitations of this study lie both in central questions and in the realisation of the study. As it was the first study of the implementation of formative peer assessment in our conditions, the study was of an exploratory nature and started with the most accessible layer of teachers' attitudes to formative peer assessment, i.e. with their perception and reflection of gained experience. To learn more about the impact that the enactment of formative peer assessment in classrooms may have on teachers' attitudes and pedagogy, observational data from their practice would be necessary. Another limitation was the relatively small size of teacher groups and the method of self-selection, which meant that we were working with teachers highly motivated for change and improvement of their teaching practice. The group was not homogenous with respect to previous experience with inquiry teaching and formative assessment, which could influence how the teachers personalised this new method of formative peer assessment in their further practice. This should be more explored. In addition, the contribution of research experience relative to other ways of professional development focused on the implementation of formative peer assessment should be investigated in a more complex design.

Conclusion

Formative assessment is one of the key strategies in science teaching. Though the idea of formativeness is, in general, accepted by teachers, the application of various methods of formative assessment is still relatively modest. Teachers themselves are mostly those who provide formative feedback; the activation of students as assessors is not very frequent. Yet the method of formative peer feedback has a potential to be very powerful, as it comprised two of five Black and Wiliam's (2009) key strategies of formative assessment (activating students as instructional resources for one another, and activating students as the owners of their own learning). To increase teacher's use of formative assessment means a complex change of his/her pedagogy, changes of the method itself does not make the difference (Heritage, 2007).

This study has shown, that immersing teachers into this relatively new method, formative peer assessment, especially when they are co-investigating their own practice (in a research project), may have a significant impact on teachers' understanding and conceptualisation of formative assessment in general.

Notes

1. The moments of contingency are described by Black and Wiliam (2009) as moments in which the evidence from learning is used for adjustments of the next steps. Though pointing on the moments may seem to be narrow focused, the considerations which are taken into account for seizing the opportunity to regulate the learning are quite broad, comprising instructional design, curriculum, epistemology, psychology, pedagogy, etc.
2. usually related to time and resource demands (worksheets, assessment tools, teacher assistant time etc.), lack of collegial support and novelty of the assessment method.
3. FP7, SiS.2012.2.2.3-1 (2013–2016): Assess Inquiry in Science, Technology and Mathematics Education (ASSIST-ME)
4. For example, primary biology unit consisted of six hours (2 per week) in which the students investigated germination. Students proposed hypotheses about the factors which influences germination, then designed an experiment. The experiment design was assessed by peers and refined. After its realisation, students summarised the outcomes and draw a conclusion. Other examples of teaching units, formative-peer-assessment activities and other materials are available at https://assistme.ku.dk.
5. The video was prepared for ASSIST-ME project by the University of Copenhagen and showed the inquiry-based lesson in which students experimentally investigated the dependence of pendulum time on its length and provided formative feedback to their peers.

Acknowledgement

We thank all pre-service teachers for their participation and cooperation.

Disclosure statement

No potential conflict of interest was reported by the authors.

Funding

This work was supported by European Commission [grant number FP7 Grant Agreement No. 321428 ASSIST-ME].

ORCID

Iva Zlabkova http://orcid.org/0000-0001-8384-4265
Jan Petr http://orcid.org/0000-0003-1484-5453
Iva Stuchlikova http://orcid.org/0000-0002-3300-8573
Lukas Rokos http://orcid.org/0000-0001-5403-1564
Alena Hospesova http://orcid.org/0000-0003-3626-6444

References

Abell, S. K., & Lederman, N. G. (2007). *Handbook of research on science education*. Mahwah, NJ: Lawrence Erlabaum Associates.

Anderson, R. (2002). Reforming science teaching. What research says about inquiry? *Journal of Science Teacher Education*, *13*(1), 1–12.

Andersson, C., & Palm, T. (2017). Characteristics of improved formative assessment practice. *Education Inquiry*, *8*(2), 104–122.

Birman, B. F., Desimone, L., Porter, A. C., & Garet, M. S. (2000). Designing professional development that works. *Educational Leadership*, *57*, 28–33.

Black, P., & Wiliam, D. (2009). Developing a theory of formative assessment. *Educational Assessment, Evaluation and Accountability*, *21*(1), 5–31.

Black, P. J., & Wiliam, D. (1998). Inside the black box: Raising standards through classroom assessment. *Phi Delta Kappan*, *80*(2), 139–148.

Blanchard, M. R., & Sampson, V. D. (2018). Fostering impactful research experiences for teachers (RETs). *Eurasia Journal of Mathematics, Science and Technology Education*, *14*(1), 447–465.

Borko, H. (2004). Professional development and teacher learning: Mapping the terrain. *Educational Researcher*, *33*(8), 3–15.

Box, C., Skoog, G., & Dabbs, J. M. (2015). A case study of teacher personal practice assessment theories and complexities of implementing formative assessment. *American Educational Research Journal*, *52*(5), 956–983.

Bryant, D. A., & Carless, R. R. (2010). Peer assessment in a test-dominated setting: Empowering, boring or facilitating examination preparation? *Educational Research for Policy and Practice*, *9* (1), 3–15.

Cooper, B., & Cowie, B. (2010). Collaborative research for assessment for learning. *Teaching and Teacher Education*, *26*(4), 979–986.

Dolin, J., & Evans, R. (2018). *Transforming assessment: Through an interplay between practice, research and policy*. Cham: Springer International Publishing.

Dolin, J., & Evans, R. H. (2013). The contribution of formative assessment and self-efficacy to inquiry learning. In M. H. Hoveid & P. Gray (Eds.), *Inquiry in science education and science teacher education* (pp. 125–145). Trondheim: Akademika Publishing.

Evans, R., Clesham, R., Dolin, J., Hošpesová, A., Birch Jensen, S., Nielsen, J. A., … Zlabkova, I. (2018). Teacher Perspectives about using formative assessment (227–248). In J. Dolin & R. Evans (Eds.), *Transforming assessment* (pp. 141–173). Cham: Springer International Publishing.

Fullan, M., & Hargreaves, A. (1992). *Teacher development and educational change*. Washington, DC: The Falmer Press.

Guskey, T. R. (2003). What makes professional development effective? *Phi Delta Kappan*, *84*(10), 748–750.

Guskey, T. R., & Yoon, K. S. (2009). What works in professional development? *Phi Delta Kappan*, *90*(7), 495–500.

Hazelkorn, E., Ryan, C., Beernaert, Y., Constantinou, C., Deca, L., Grangeat, M., … Welzel-Breuer, M. (2015). *Science education for Responsible Citizenship (No. EUR 26893)*. Brussels: European Commission - Research and Innovation.

Heinz, J., Enghag, M., Stuchlíková, I., Cakmakci, G., Peleg, R., & Baram-Tsabari, A. (2018). Impact of initiatives to implement science inquiry: A comparative study of the Turkish, Israeli, Swedish and Czech science education systems. *Cultural Studies of Science Education*, *13*(2), 607–615.

Heitink, M. C., Van der Kleij, F. M., Veldkamp, B. P., Schildkamp, K., & Kippers, W. B. (2016). A systematic review of prerequisites for implementing assessment for learning in classroom practice. *Educational Research Review*, *17*, 50–62.

Heritage, M. (2007). Formative assessment: What do teachers need to know and do? *Phi Delta Kappan*, *89*(2), 140–145.

Kingston, N., & Nash, B. (2011). Formative assessment: A meta-analysis and a call for research. *Educational Measurement: Issues and Practice*, *30*(4), 28–37.

Popping, R. (1988). On agreement indices for nominal data. In W. E. Saris & I. N. Gallhofer (Eds.), *Sociometric research: Volume 1, data collection and scaling* (pp. 90–105). New York: St. Martin's Press.

Reavey, P. (2018). *Visual methods in psychology: Using and interpreting images in qualitative research*. Kobenhaven: Nota.

Rocard, M., Csermely, P., Jorde, D., Lenzen, D., Walberg-Henriksson, H., & Hemmo, V. (2007). *Science education now: A renewed pedagogy for the future of Europe* (1st ed.). Luxembourg:

Office for Official Publications of the European Communities. Retrieved from http://ec.europa.eu/research/science-society/document_library/pdf_06/report-rocard-on-science-education_en.pdf

Šeďová, K., Švaříček, R., & Šalamounová, Z. (2012). *Komunikace ve školní třídě* [*Communication in the classroom*]. Praha: Portál.

Shepard, L. A. (2000). The role of assessment in a learning culture. *Educational Researcher, 29*(7), 4–14.

Shepard, L. A. (2005). Linking formative assessment to Scaffolding. *Educational Leadership, 63*(3), 66–70.

Strijbos, J. W., & Sluijsmans, D. (2010). Unravelling peer assessment: Methodological, functional, and conceptual developments. *Learning and Instruction, 20*(4), 265–269.

Strijbos, J., & Wichmann, A. (2018). Promoting learning by leveraging the collaborative nature of formative peer assessment with instructional scaffolds. *European Journal of Psychology of Education, 33*(1), 1–9.

Stuchlíková, I., Petr, J., & Papáček, M. (2013). Inquiry-based teaching and future teachers' attitudes towards it. In M. H. Hoveid & P. Grey (Eds.), *Inquiry in science education and science Teacher education. Research on teaching and learning through inquiry based approaches in science (teacher) education* (pp. 167–189). Trondheim: Akademika Publishing.

Sun, X., & Yang, Y. (2018). *Generalized McNemar's test for homogeneity of the marginal distributions.* Retrieved from http://www2.sas.com/proceedings/forum2008/382-2008.pdf

Sunal, D., Hodges, J., Sunal, C., Whitaker, K., Freeman, L., Edwards, L., et al. (2001). Teaching science in higher education: Faculty professional development and barriers to change. *School Science and Mathematics, 101*(5), 246–257.

Topping, K. J. (1998). Peer assessment between students in colleges and universities. *Review of Educational Research, 68*(3), 249–276.

Topping, K. J. (2009). Peer assessment. *Theory Into Practice, 48*(1), 20–27.

Van den Bossche, P., Gijselaers, W., Segers, M., & Kirschner, P. A. (2006). Social and cognitive factors driving teamwork in collaborative learning environments: Team learning beliefs and behaviors. *Small Group Research, 37*(5), 490–521.

Van Gennip, N. A. E., Segers, M. S. R., & Tillema, H. H. (2009). Peer assessment for learning from a social perspective: The influence of interpersonal variables and structural features. *Educational Research Review, 4*(1), 41–54.

Van Zundert, M., Sluijsmans, D. M. A., & Van Merriënboer, J. J. G. (2010). Effective peer assessment processes: Research findings and future directions. *Learning and Instruction, 20*(4), 270–279.

Wagner, R. F. (2003). Clinical case formulation in the research program "Subjective Theories". Patients with chronic diseases. *European Journal of Psychological Assessment, 19*(3), 185–194.

Wilcox, J., Kruse, J. V., & Clough, M. P. (2015). Teaching science through inquiry: Seven common myths about this time-honored approach. *The Science Teacher, 82*, 62–67.

Windschitl, M. (2003). Inquiry projects in science teacher education: What can investigative experiences reveal about teacher thinking and eventual classroom practice? *Science Education, 87*(1), 112–143.

The what, when & how factors: reflections on classroom assessment in the service of inquiry

Bronwen Cowie and Christine Harrison

ABSTRACT
This paper provides both a summary and plenary paper for a Special Issue of papers that build on the findings of a research project that investigated formative and summative assessment methods to support and to improve inquiry-based science teaching and learning. The paper aims to unravel some of the complexity that underpins classroom assessment practices as teachers bring change to their pedagogical approaches, taking into consideration both the variability and variation in interpretation of both processes by STEM teachers across eight countries. By engaging with research processes, in collaborative working groups in each country, teachers were better equipped to face the practical challenges they encountered in their various contexts, strengthening both their processes of critical reasoning and exercise of professional judgement. We draw from this interweaving of connected professional ideas and nested practices, a consideration of those factors which seem most pertinent in terms of formality, temporality, response and desired outcomes. As such, the paper provides a directive that highlights how the development of inquiry competences and the skills to negotiate productive pathways through an inquiry activity can be informed, developed and enhanced by formative assessment. Such developments require pedagogic moves that allow student agency while fostering advancement through guided feedback.

Introduction

Classroom assessment and STEM inquiry are a focus for policy, research and practice worldwide. The papers in this special issue explore the interaction of these two agendas drawing on the close collaborative work between researchers and teachers as part of the Assess Inquiry in Science, Technology and Mathematics Education (ASSIST-ME) project. This was an eight-country research project that investigated formative and summative assessment methods to support and to improve STEM inquiry learning within eight European countries. Based on an analysis of what is known about summative and formative assessment of the knowledge, skills and attitudes related to key STEM competences, and an analysis of the educational systems of the eight countries involved, the project designed a range of assessment approaches for STEM inquiry in primary and secondary classrooms. These were customised by Local Working Groups (LWGs) in each country to accommodate their particular educational contexts. In this paper we analyse the individual and combined contributions of the papers in the special issue, connecting these with the reported findings of the ASSIST-ME project book *Transforming Assessment: Through an interplay between practice, research and policy* (Dolin & Evans, 2018), and to main literature in the fields of inquiry and assessment. In this way, we are able to explore practices, outcomes and challenges across a variety of contexts to provide insights into both the uptake and the development of classroom assessment practices in inquiry settings.

The ideas we put forward and the connections we make arise from a stream of conversations over more than a two-year period about the approach, workings and findings of the ASSIST-ME project as we puzzled over the complexities of introducing changes in assessment practices alongside developing inquiry learning in STEM. We were in a unique position to ponder over this matter since one of us had taken part in the project and was familiar with the data and analysis and the other looked on from the perimeter making sense of the interpretations and the challenges encountered. Specifically, we consider how formative assessment activities that enhance STEM inquiry operate across the dimensions of formality, form and function; and time, temporality and differential support for diverse learning outcomes. We also consider how assessment can be both a private and or public activity that makes up people, predicated through the response to feedback and its effect on self-esteem and efficacy.

We acknowledge that while recent OECD publications stress the importance of formative assessment and its relationship with summative assessment (e.g. Looney, 2011; Lucas et al., 2013; OECD, 2013), they also identify that classroom assessment in many countries still seems to be dominated by summative assessment, and that teacher classroom assessment practice is largely dominated by accountability pressures from school leaders and governments. In such environments, teachers often form deeply held pedagogical beliefs about assessment as a tool for teacher use and accountability, rather than as a means of guidance and support for learning and this can prevent formative assessment becoming a regular and valued part of teaching. To ensure teachers reconcile competing assessment tensions, there is a need to challenge teachers' conceptions of assessment as well as to provide opportunities for professional learning about formative practice in STEM classrooms (Correia & Harrison, 2019). In this paper, we also draw attention to the variability in teacher practice within and across countries, and also within and across different STEM disciplines and phases of education. We conclude by outlining the next steps for research and practice.

A sociocultural lens for reflection on and refraction of classroom assessment

Teaching, learning and assessment are understood as cumulative multidimensional activities that are enacted in action and discourse across different spaces and over different scales of time. The relationship between teaching, learning and assessment is problematic as each aspect is precariously balanced in respect to the others. This paper is underpinned by a sociocultural view of learning and its assessment (Gipps, 1999; Moss et al., 2008;

Shepard, 2000; Shepard et al., 2016). This view acknowledges the social, relational, material and temporal dimensions of learning and construes learning as a process of identity change and development. More specifically, it acknowledges the extent to which interaction and action are contingent on and entangled with the people and established patterns of participation in a context and the conceptual and material/virtual resources that happen to be at hand (Cowie et al., 2013). Moss (2008) sums up our understanding in her claim that learning, assessment and the exercise of agency cannot be separated from how the classroom learning environment is resourced with 'knowledgeable people, material and conceptual tools, norms and routines, and evolving information about learning' (p. 254).

Formative assessment as part of inquiry opens up opportunities for students to take charge of their own learning journey (Harlen, 2011). Inquiry includes ample opportunities for students to pose, explore and test out their ideas. This exercise of agency can build student confidence and competencies, yet, at the same time, next steps are productively guided and constrained evidence generated alongside and through these activities. Feedback can come from a range of sources – peers, observations, discussions and inferences, the teacher and other knowledge resources such as the internet. These multiple sources of feedback provide a rich means of guidance about more and less productive lines of inquiry. In particular, students benefit from teacher attention to the longer-term fruitfulness of some of their ideas compared to others. So, while the pathway and goals may be clear at the start, some events will cause teachers, and possibly students, to reflect on their initial assumptions, while others shape and modify the dimensions of the progression ahead of them. This suggests that inquiry learning can never be assumed as a linear trajectory but rather an evolving meandering path ranging across a changing educational landscape with reflection on experience adding to the meaning of experience and helping to direct the course of subsequent experience (Dewey, 1916).

Formative assessment practice within inquiry

Teachers and researchers in the ASSIST-ME study investigated the use of many of the commonly recommended formative assessment strategies within the context of science inquiry tasks –. assessment on-the-fly, structured assessment dialogue, peer assessment, and written assessment feedback. These methods were tested in primary and secondary schools in different educational cultures and contexts in Europe in order to analyse the conditions that support or undermine the uptake of formative assessment related to inquiry processes. Looking across the papers, it is possible to discern that these can be analysed in terms of the following linked aspects:

Formality, form and function

Time, temporality and differential support for diverse learning outcomes

A private or public activity that makes up people

Each of these aspects encompasses an array of possibilities for teacher focus and action and each comes with consequences for students' day-to-day educational experiences and identity trajectories. Hence, the final aspect we address is that of the variation in teacher focus and action. While our expectations are that it is difficult to justify generalisable trends because of the multiple relevant variables within education in diverse settings, the growth in useful knowledge and confidence among the project teachers points to various indicators for further productive development within national contexts.

Formality, form and function

When assessment is understood as a social practice, classroom assessments can usefully be understood in terms of their formality (Shavelson et al., 2008) and linked to purpose and function. Teacher classroom assessment can be undertaken for both formative or summative purposes (Dolin, Black, et al., 2018) but a common misconception is that it is the task rather than the teacher's action that determines this purpose. Recognition that the same assessment activity can accomplish multiple purposes has focused attention on the function of an assessment, that is on the actual consequences of teacher assessment actions for students and their learning, broadly conceptualised (Harrison et al., 2018; Hickey, 2015). We adopt this focus in this paper.

With formal and structured formative assessment, teachers generally plan activities to elicit specific information about curriculum learning outcomes from the class as a whole. In Torrance and Pryor's (2001) terms these activities are a convergent enactment of formative assessment; they converge on the learning the teacher is intending. Informal, spontaneous or contingent formative assessment tends to be grounded in teacher observation and or interaction with students (Harrison, 2006), with these actions essentially indivisible from teaching (Moss, 2008). In this case, teachers are more likely focusing on what a student is actually doing and thinking, and hence undertaking what Torrance and Pryor refer to as divergent formative assessment. Both planned and contingent, convergent and divergent formative assessments are needed to ensure students make progress in ways that build on their knowledge and competences which help them understand the subject discipline and achieve a balance between these different aspects. Insights into both kinds of formative assessment can be found in the papers in the special issue.

A number of researchers have put forward descriptions of the process that teachers go through as they engage in informal formative assessment on-the-fly (e.g. Cowie & Bell, 1999; Harrison et al., 2018; Ruiz-Primo & Furtak, 2007; Torrance & Pryor, 2001). The ASSIST-ME team explored the use of Ruiz-Primo and Furtak's (2007) model of ESRU (Eliciting, Student response, Recognising and Using) to analyse teacher informal assessment practices in their different countries. Their findings provide evidence that teachers work through both complete and incomplete ESRU cycles. They illustrate that sometimes teachers act to use evidence quickly and at other times they accumulate evidence by moving through a number of partial cycles, collecting ideas from a number of students, before taking action (Nieminen, Hähkiöniemi and Viiri, 2020; Rached & Grangeat, 2020; Sezen-Barrie & Kelly, 2017). Rached and Grangeat (2020) alert us to the role students can play in eliciting, recognising and using information from peers, thereby positioning them as agentic. They also add consideration of the role of non-verbal contributions such as gestures, illustrations and diagrams alongside dialogue in providing assessment evidence. These expansions are important in positioning students as active in the assessment process and in acknowledging the multimodal nature of making sense of and using scientific concepts (Cowie et al., 2013; Lemke, 2000a). This said, as Sezen-Barrie and Kelly (2017) reported, even when teachers incorporate student responses into their classroom dialogue, student contributions do not necessarily lead to a change in the direction of the dialogue but rather are selected only by the teachers if they fit with the argument the teacher is making, and rejected or ignored if not. These classroom actions are perhaps shaped by the teacher's focus on pre-determined learning goals or targets. Sezen-Barrie and Kelly's findings echo those of Torrance and Pryor (2001) in highlighting the challenge for teachers of balancing being responsive to student ideas and ensuring they develop sophisticated disciplinary understandings, that is in managing the tension between a divergent and a convergent formative focus (See also Cowie et al., 2018). This is all the more challenging during student directed inquiries, where students take a more active role in decision-making.

Nieminen, Hähkiöniemi and Viiri (2020) provide a nuanced interpretation of the processes two teachers use to elicit and use information. Their analysis extends beyond a focus on dialogic moves to consider the overall outcome and tone of classroom episodes. This involved them looking for patterns of elicitation and the consequent teacher use of evidence following an overt elicitation move *and* looking for an elicitation move when teacher information use could be inferred. They identified five forms of eliciting: by open and closed questioning, by listening and observing, and by recalling for student commentary previous findings and four main categories for use: checking or accepting thinking, correcting or developing thinking, redirecting thinking and reviewing thinking. It is interesting to note that these categories include actions consistent with both convergent and divergent formative assessment. Also of note is Nieminen, Hähkiöniemi and Viiri's comment that it was often difficult to distinguish between teacher elicitation and teacher information use, in part because teachers drew on evidence generated previously, both within that specific lesson and in previous lessons. They also recognised that students could and did initiate interactions around ideas, suggesting a more shared responsibility and direction for learning in these classrooms.

Students asking questions and seeking help is not always considered within the formative assessment literature but we propose that because help-seeking requires students to identify when and what assistance they need, and from what or whom, it can be seen as a form of student-initiated formative self-assessment action (Boud & Molloy, 2013; Cowie, 1998). Within an inquiry the help sought could be to progress a task may entail choosing a strategy, identifying a suitable source of evidence, seeking guidance to process evidence, and or seeking teacher/peer advice (Harrison, 2014). As such, help-seeking is framed by a learner's awareness about the context in respect to persons, strategies, evidence, actions and goals that may be of help (Fletcher, 2018; Moss, 2008). Studies on what is required for students to use feedback formatively highlight its entangled relational-dialogical-affective nature (Carless, 2013; Ellegaard et al., 2017; Esterhazy & Damsa, 2017; Evans, 2013). Feedback as an ongoing dialogue between teachers and students has been shown to support student use of formative feedback (e.g. Nicol, 2010; Shute, 2008; Torrance & Pryor, 2001). Written feedback in particular needs to serve as an invitation to dialogue; it needs to communicate respect for the student as a learner, to position them as active and to inspire thought, curiosity, and or wondering (Brookhart, 2008). Students report that feedback in the form of open, wondering or leading questions offers them a 'thinking device' (Dysthe, 1996, p. 395; Cowie, 2005). Feedback of this kind is at the most formative end of the scale (Dolin & Evans, 2018) and guides students towards self-directed learning.

Time and temporality

The dynamic nature of formative assessment in inquiry classrooms is worthy of further exploration. A number of scholars have highlighted the role of time and temporality in teaching and in learning more generally (e.g. Lemke, 2000b; Mercer, 2008). These scholars point out that learning happens over time and multiple timescales. Adding complexity, as Mercer (1995) points out, 'all conversations have a history and a future and take place between particular people in a specific place and time' (p. 30). Depending on the focus for learning and the teacher's judgement about the appropriate time and way to progress student learning formative assessment cycles can be accomplished over shorter or longer timescales (Cowie & Moreland, 2015; Harrison, 2015; Leahy et al., 2005). With on-the-fly formative assessment teachers and students make sense of and use information in real time and in the context of its generation (Heritage, 2013), often employing resources from the context as part of their feedback interactions (Cowie et al., 2013). While these in the moment responses are central to responsive formative assessment a caveat is that teachers can be challenged to include and respond to *all* their students within a timely manner (Cowie, 2015).

In this Special Issue, the paper by Rached and Grangeat (2020) focuses on how an argumentation approach to formative assessment can make student thinking visible. In their study, they analyse the classroom talk during the planning of investigations and presentation of ideas by students to their peers. They classify the conversations both in terms of the formative purposes using Ruis-Primo and Furtak's ESRU methodology and also by classifying utterances as epistemological, conceptual or social (Duschl, 2008). In this way, teachers not only pick up on the dynamics of the classroom talk but also the interplay between student's current levels of inquiry thinking and the aspirational goals that they are trying to engender in their classroom. In many ways, rather than attempting to 'close the gap' between student understanding and teacher goals, these conversations begin to form the bridge that supports student learning and progress.

The ASSIST-ME studies add to our understanding of temporality of formative assessment as situated in both the school setting and what is to be learned. For example, Grob et al. (2019) found that primary school teachers generally implemented formative assessment over a shorter time span than upper secondary school teachers. Different competences were assessed over different cycle lengths with domain-specific competencies assessed through shorter cycles than transversal competences at both school levels. With transversal competences, the transfer to subsequent activities and situations was more frequent enabling teachers and students to transfer ideas and develop them in new contexts Primary teachers tended to offer students opportunities to use feedback to revise their original artefact whereas upper secondary school teachers required students to transfer ideas to a similar subsequent task or situation. Primary could be seen to focus more deepening and developing ideas in a particular context consistent with a convergent approach whereas secondary teachers focused more on application suggesting a more divergent focus. At both school levels, albeit in different ways, evidence of competence was collected over several occasions and so judgements and consequential formative action were driven not simply by a single student performance but by a growing picture of that competence distilled from different inquiry contexts (Dolin & Evans, 2018). This growing conceptualisation of student competence through the build-up evidence enables teachers to qualify and calibrate their expectations both for individual students and for age-related performance, building teacher assessment capability (Sadler, 1989; 2010). This process helps build up the layers of evidence through which teachers come to make judgements about future student performance.

This in turn suggests a complex interaction between formative and summative assessment where the teacher notices a spectre of previous performance within the current assessment episode. At the same time, it is important to acknowledge that teachers can and do accumulate and aggregate evidence on their students' ideas and inclinations over time and contexts. Their responses may be prompted by a combination of evidence presented in the moment and their recollection and synthesis of evidence from those specific learners and or from other students at a similar point in their learning and or in a similar context. They can also take time to consider their feedback responses and they can revisit student ideas on multiple occasions (Cowie, 2005). This layered effect of evidence consideration guides and shapes any ensuing dialogue and informs the choices offered by teachers and also those taken up by students during any inquiry. These possibilities suggest any interpretation of action based on the immediate conversational flow may provide too simple an analysis of the formative cycle.

This is an important consideration in teacher formative assessment practice given some valued inquiry learning outcomes will be relevant and an essential focus across the timescale of a lesson or activity, whilst others are an appropriate and important focus over the course of an inquiry, or even the school year. Scott et al. (2011) argue that pedagogical link making is a key aspect of teaching and learning scientific conceptual knowledge. Seen this way, scaffolding and feedback to build student appreciation of continuity and connection between ideas, tasks, and lessons is a legitimate focus for formative assessment (Cowie et al., 2013). This is essential when the aim is for students to understand and appreciate the disciplinary narrative that overarches and connects the learning goals in the short, medium and longer term. Also, the longer-term goals associated with productive inquiry in science such as the capacity to articulate and to consider a range of evidence and options for action means that teachers can usefully attend to student understanding of the social nature of learning as joint and shared responsibility (Cowie et al., 2018; Cowie & Moreland, 2015; Marshall & Drummond, 2006).

A public-private activity that makes up people

Findings from the ASSIST-ME project bring into sharp relief the dynamics of the public-private nature of classroom-based assessment. Summative assessment has traditionally been associated with the testing of individuals – students sit by themselves and record their answers to written questions. Typically, no negotiation about the selection, meaning or appropriateness of questions or the meaning or appropriateness of student answers is permitted. Similarly, any feedback – marks and or grades – tends to be private, communicated to the individual. Students have the choice to share or not this information. On the other hand, whole-class dialogue as a teaching and formative assessment strategy is a very public activity, where individual performance is noted and considered by all participants. In the context of classrooms, even group-teacher, individual student-teacher, and student-student interactions are open to being overheard and or deliberately scrutinised by others – the teacher and other students. This comes with implications for student identity – their sense of self and their reputations within a class as knowers and learners (Cowie, 2005; Crooks, 1988; Gipps, 1999). Paraphrasing Stobart (2008) even informal interactions around ideas serve to 'make up people'. Productive participation in STEM inquiries involves students both reflecting on and taking responsibility publically for the decisions they make about the next steps when the outcome is often ambiguous for them.

Sadler (1989) was very clear that the ultimate goal of formative assessment is that students develop the capacity and confidence to monitor and progress their own learning – that is they join the guild of those who can articulate and monitor their developing understanding. Within an inquiry students need to continuously monitor their ideas and actions if they are to achieve their goals and to experience a 'science identity'. This requires students to have both the confidence to express their ideas in the various arenas and modes available during an inquiry and the resilience and forethought to compare their thinking and plans with the ideas and evidence generated by others in the classroom. While there are opportunities for students to recognise productive and less fruitful ways forward, they often step back from debates where they realise other ideas are being valued. This leads to a tendency to agree with the consensus and may persuade students to accept ideas without questioning the justification. In addition, classroom discussion often generates evidence that renders students more sensitive to ideas that match with the teacher's disciplinary aims. Hence, the dynamics of the classroom dialogue may shift the direction of thinking or simply persuade students to refrain and ignore evidence.

The development of students as assessors – both peer and self is central to the effective conduct of student-led inquiry. Students taking responsibility for assessing their learning and that of their peers shifts the social and intellectual authority dynamic within a classroom. For this to happen, teachers need to share their authority for and over knowledge in a way that ensures students become confident, willing and able to self-assess and provide useful feedback (Heritage & Heritage, 2013). At the same time, teachers need to believe in students' commitment to steering inquiry and taking responsibility for their learning via formative action. Peer assessment opens up student ideas and works to in-depth scrutiny that has social/relational-conceptual-motivational implications for them as assessors and assessees (Zlabkova et al., 2020). A teacher in the paper by Zlabkova et al. (2020) addressed the challenge of this by developing students as 'smart prompters' for their peers. Important in this approach, students were not to provide direct advice, rather they needed to pose 'smart' questions to assist peers to think through what might be productive next steps. This approach is consistent with what is known about feedback that students are likely to use, and it positions peers as both agentic and discerning. Hence, it supports student identities that include those of knowledge generator and assessor.

Folding back to the earlier discussion of the role of time and the temporality of learning it is important to acknowledge that student response to classroom assessments is influenced by their previous experiences of it. Students 'carry with them the totality of their experiences of learning and being assessed and this influences their responses at any given time, and context' (Boud, 1995, p. 36). Ecclestone and Pryor (2003) propose that the beliefs, attitudes, understanding and practices students develop around assessment through their experience of an assessment regime constitute their assessment career, we suggest they constitute their identity as assessor and assessee (Price et al., 2011); students need time to experience the value of self and peer assessment and feedback as a tool for strengthening and reconfiguring their identities as science inquirers and assessors.

Variation, variability and change in teacher practice

Working across eight countries, each with their own history of assessment and inquiry, highlighted the complexity and diversity of approaches that can evolve within and across teacher practice. By taking on a more formative approach, the teachers came to see their role in the classroom and their relationship with their students differently and this led to them reshaping their professional identity. By fostering professional learning, the ASSIST-ME project encouraged the teachers to become researchers of their own practice. This enabled teachers to take control of their own learning journey and to make the changes they saw fit, while at the same time, gaining support for and legitimacy from these actions from the professional community established in their Local Working Group (LWG) (Harrison, 2013). Affirmation from other teachers and the research assisted them in reconstructing their pedagogical identity as an assessor (Harrison, 2013). Taken together, the papers provide an example of a theory of action that spans curriculum-pedagogy-assessment-professional learning (Brookhart, 2018) as a way of supporting teachers to make the knowledge, practice and identity changes needed for classrooms to become contexts where teachers and students are partners in and for learning (Shepard et al., 2018).

Teachers attempting to change their pedagogic practices at the same time as their assessment practices are especially challenging (Harrison, 2013). They need to reconcile and resolve any tensions between their understanding of inquiry and of formative assessment before they could make changes to establish a new norm from the meld of previous practices. Within the ASSIST-ME project, teachers from different countries had different views and experiences of formative assessment practices, but even those more experienced in formative practice were taxed when introducing these into STEM inquiry learning. Their responses emphasised that teachers need a variety of living examples of implementation, preferably from teachers with whom they can identify and from whom they can both derive conviction and confidence (Black et al., 2003; Black & Harrison, 2004). Practice-based evidence (Gittner & Harrison, 2019) helped provide a vision of how inquiry supported by formative assessment could be which helped teachers conceptualise how it might work in their context. Teachers needed to recognise the elements and value/s of both innovations and consider the implications these might involve for their students as part of bringing together and combining aspects of the innovations within their current ways of teaching to bring about change. The LWGs on the ASSIST-ME project provided platforms for the articulation of and public reflection on practice with like-minded professional colleagues that provided both an impetus for pursuing change as well as support to hone ideas in a non-judgemental manner. They ensured teacher professional knowledge was made public in a form that enabled it to be accumulated, shared and continually endorsed and enhanced.

It is of note that each country's Local Working Group (LWG) focused their main attention on different parts of the inquiry cycle and prioritised different aspects of formative assessment as part of determining 'where to start' within their localised framing (Cowie et al., 2018; Richards et al., 2020). For the Swiss LWGs (Grob et al., 2019), they approached the professional support by researchers and teachers working together collaboratively to identify an inquiry and decide which inquiry activities to do then select which form of formative action seemed to fit most appropriately with the range of activities in the topic. In this way, the LWGs helped teachers recognise the opportunities for different types of formative assessment; for example, whether to provide teacher or peer feedback as part of the formative drive in an activity. For the French LWGs (Rached & Grangeat, 2020), the focus was more on how formative assessment supported the planning stages of an investigation through considering the conversational moves that might occur between teachers and students, and also within whole-class discussions as students planned, presented, modified and tested out their inquiry approaches. This enabled teachers to be ready and responsive to student ideas and actions.

Various tools were developed by the LWGs to support the teacher change process. Rached and Grangeat (2020) explain how they constructed 'progression charts' which helped teachers communicate to students four different levels of achievement within specific inquiry competencies. Teachers and students used the charts to interpret and recognise the development of inquiry competencies across a number of contexts, and over time the progression chart framework built 'common knowledge' (Edwards & Mercer, 1987). One of the main difficulties in this area is the decisions about appropriate granularity or grain size for progressions for them to enable students to engage in peer assessment, a matter that is debated in relation to the use of learning progressions to guide teacher assessment (e.g. Duschl, 2008). Through an iterative process of sharing and acting on practice-based evidence the French LWG developed a series of progression charts that were locally and situationally meaningful. Their development process provides an example of how groups can exercise consistency of principle rather than uniformity of practice (Harrison & Howard, 2009) by drawing on local expertise and evidence/ experience as a resource.

The papers in this Special Issue offer insights into changes in teacher confidence and self-efficacy consequent to teacher participation in the ASSIST-ME professional learning programme. This was evident both in the measures that were taken to track self-efficacy through teacher questionnaires at the start and end of the periods when LWGs were working on designing, teaching and developing inquiry teaching units, and for the same group 20 months later (Zlabkova et al., 2020). It was also informed by the interview data and the dialogue captured in the LWG meetings, and, in the case of the French LWGs by video footage of the lessons. At the same time, it provided evidence that teacher self-efficacy was informed and nested within the development of localised collective efficacy (Tschannen-Moran & Barr, 2004), in this case collective efficacy across the group of collaborating teachers. The increase in collective efficacy and common knowledge within the various LWGs, supported the sharing of ideas, classroom experimentation and receptivity to feedback from others. These various aspects were key to teachers' willingness and capacity to explore the productive interplay between formative assessment and instructional support for student inquiry, especially more open inquiries. While this led to some commonality within a LWG, across the levels of teacher experience and across the different country contexts a wide range of practices evolved testifying to the need to and value of customised professional learning opportunities.

The papers attest to the centrality and complexity of teacher knowledge involved in deciding how and when to respond to evidence arising from student actions to meet the multitude of demands arising as part of any STEM inquiry. Teachers need curriculum and assessment knowledge as part of their subject-specific professional knowledge and they need subject and pedagogical knowledge, including pedagogical content knowledge, as part of their assessment knowledge (Ropohl & Rönnebeck, 2019). The mutualistic dynamism of these knowledges lies at the core of what teachers do when they become effective in STEM inquiry classrooms. It is teacher understanding of possible student alternative conceptions and likely pathways for knowledge development that enables them to take informed action (Ropohl & Rönnebeck, 2019; see also Coffey et al., 2011). Teachers' understanding of whether and how a particular idea may be important in the longer term and what a good performance in their specific short-term learning goals looks like, and also what it does not look like, is also essential to productive formative action (Heritage et al., 2009).

In their reflections of their professional journey, ASSIST-ME project teachers identified the far-reaching benefits of the LWGs approach, the positive engagement of their students and the enhancement of their (teacher) capability to capture, interpret and act on evidence of learning in inquiry contexts (Dolin & Evans, 2018). The teachers also highlighted the demands of this approach, while still trying to satisfy the current teaching and assessment requirements of their particular cultural context. The teachers commented on the time, effort and persistence required to incorporate inquiry-based teaching approaches and different assessment tools within their current practice. In some instances, this approach challenged existing teacher perceptions of what inquiry learning looks like (autonomous, practical, authentic, collaborative) and the implications this has for assessment evidence (unpredictable, difficult to document on a whole-class basis, difficult to interpret on an individual student basis). There was a growing appreciation that such practices require the use of qualitative judgement (Sadler, 1989), which often involves the use of multiple criteria across different inquiry contexts. Importantly, as Sadler notes, aspects that are salient may not be able to be specified in advance or be uniformly applicable to different student responses.

It was a key aspect of the ASSIST-ME project to provide a research base on effective uptake of formative assessment for inquiry-based, competence-oriented STEM education. However, the project partners were extremely aware that these processes all took place in educational systems dominated by high stakes summative testing. So, while the LWGs could develop new assessment methods for formative use, the implementation of these methods and the realisation of their true potential was to some extent limited by the political realities, that gave high priority to the summative use of assessment. Therefore, the project partners set up a second aim: To use the research base to inform policymakers and other relevant stakeholders to enable them to consider how they might reconcile formative and summative assessment tensions within future policy and practice. These policy-oriented aspects were dealt with in National Stakeholder Panels (NSPs) established in each country. The members of these NSPs were representatives from different stakeholder groups influencing educational decisions in the country (Dolin, Bruun, et al., 2018). The NSPs discussed matters such as how to secure meaningful communication between teachers, researchers and stakeholders, the need for teacher professional development in assessment, and the functioning of the national assessment system. The NSP national project partners established some general recommendations for how existing assessment systems might be modified to enable formative assessment to function effectively in STEM classrooms.

Concluding comments

The set of papers in this special issue make an important contribution to our understanding of classroom assessment in support of inquiry-based STEM education, highlighting the importance of the intersection between formative assessment and inquiry processes. They highlight how the development of inquiry competences and the skills to negotiate productive pathways through an inquiry activity can be informed and enhanced by formative assessment. They recognise that informal formative assessment activities, whether generated by the teacher or the students, need to be coherent between each other and with the characteristics of the instruction and the curriculum. They also need to be comprehensive in that they generate evidence over contexts and time in ways that support teacher instructional decisions and inform student self-monitoring and action.

How teachers respond to assessment evidence that arises from classroom activities is governed by a mixture of observation, interaction and judgment that is informed by beliefs about learning and assessment purposes. The STEM teachers that are the focus of these papers were supported in formative assessment processes by the development of new routines, tools and or practices that were devised to strengthen classroom interactions and inform their judgements about student learning. These new practices were explored as part of understanding the demands on teachers as they were seeking to enact new practices in classroom assessment that also aligned with inquiry instruction and also to understand the degree and direction that professional learning opportunities might take. The papers present useful insights for the field in how to cultivate a supportive environment for teacher and student reflective engagement in and enactment of formative assessment in the context of inquiry. These are complex and non-linear processes that vary in formality, form and function, and can be enacted and experienced along multiple time scales in the context of learning the different knowledges, practices and dispositions that comprise science inquiry. Significantly, the papers provide an insight into how formative assessment makes up and can remake both teachers and students as science inquirers and as assessors within the dynamics of an inquiry classroom informed by teacher and student engagement in classroom discussions, observation, experimentation and formative assessment conversations.

The ASSIST-ME project suggests that investing in professional learning, such as providing guidelines and tools to facilitate formative assessment practice, and by encouraging innovation and creating opportunities for teachers to experiment with assessments in inquiry contexts needs to form a key part of future development of STEM inquiry-based learning. The papers provide support for the value of teacher-researcher collaboration coupled with the introduction of material, virtual and educational assessment tools (rubrics, learning progressions, knowledge about effective feedback) to assist teachers to understand and implement formative assessment aligned with summative assessment. Within ASSIST-ME, teachers and researchers collaborated in 'local working groups' to plan, enact and reflect on formative practice in support of inquiry over three years. Particular attention was given to developing teachers' understanding of science inquiry as a complex activity and how the various inquiry processes intersected with formative and summative approaches to assessment. The meetings between teachers and researchers provided opportunities for teachers to share, reflect and re-interpret evidence arising from their classroom assessment practices and the changes they were making in instructional and assessment approaches.

Many of the teachers were of the view that they would not have explored and experimented with formative assessment without researcher support and affirmation from the other project teachers. More significantly, they concluded the use of formative assessment can only be constructively learned in a situated way. Through professional dialogue and sharing their emerging thinking about inquiry learning and associated assessment practices the teachers were able to articulate and justify what and why certain classroom decisions were made. This process helped the teachers strengthen their understanding of what constitutes effective assessment for inquiry learning. The teachers' commentary provided support for direct attention to fostering collective efficacy as a shared commitment to and belief in being able to accomplish a substantial change in classroom practice. At the same time the variability and variation in the ASSIST-ME country and teacher contexts, practice and developments direct our attention to the need to better understand the 'contours of variation' (Brky, 2015, p. 471) within and across the factors likely to contribute teacher change so that we can achieve better outcomes for all learners more reliably and at scale.

This concluding article, drawing both on the special issue articles and the whole project, provides some important avenues for further research. It proposes a new framework for analysing formative assessment and summative assessment practices, through the use of (i) Formality, form and function, (ii) Time, temporality and differential support for diverse learning outcomes and (iii) A private or public activity that makes up people. It highlights two new questions: How do teachers and students cope with the public-private nature of classroom-based assessment? and How can teachers build student appreciation of continuity and connection between ideas, activities, and lessons? The articles in this issue illustrate how researcher-teacher collaboration can influence the way teachers teach their classes within the given educational frame. Nevertheless, it became clear through the progress of the project that national assessment systems need to change in order to enable teachers to introduce and develop new teaching methods with an emphasis on formative processes. The establishment of the National Stakeholder Panels in each project country showed one way to affect the policy system through linking teachers, researchers and policymakers.

Disclosure statement

No potential conflict of interest was reported by the author(s).

ORCID

Bronwen Cowie http://orcid.org/0000-0003-3578-0791

References

Black, P., & Harrison, C. (2004). *Science inside the Black Box: Assessment for learning in the science classroom*. NfER Nelson.

Black, P., Harrison, C., Marshall, B., Wiliam, D., & Lee, C. (2003). *Assessment for learning: Putting it into practice.* Open University Press.

Boud, D. (1995). Assessment and learning: Contradictory or complementary? In P. Knight (Ed.), *Assessment for learning in higher education* (pp. 35–48). Kogan Page.

Boud, D., & Molloy, E. (2013). Rethinking models of feedback for learning: The challenge of design. *Assessment & Evaluation in Higher Education, 38*(6), 698–712. https://doi.org/10.1080/02602938.2012.691462

Brky, A. (2015). Accelerating how we learn to improve. *Educational Researcher, 44*(9), 467–477. https://doi.org/10.3102/0013189X15621543

Brookhart, S. (2008). *How to give effective feedback to your students.* ASCD.

Brookhart, S. (2018). Learning is the primary source of coherence in assessment. *Educational Measurement: Issues and Practice, 37*(1), 35–38. https://doi.org/10.1111/emip.12190

Carless, D. (2013). Trust and its role in facilitating dialogic feedback. In D. Boud, & E. Molloy (Eds.), *Feedback in higher and professional education: Understanding it and doing it well* (pp. 90–103). Routledge.

Coffey, J. E., Hammer, D., Levin, D. M., & Grant, T. (2011). The missing disciplinary substance of formative assessment. *Journal of Research in Science Teaching, 48*(10), 1109–1136. https://doi.org/10.1002/tea.20440

Cowie, B. (1998). The role of questions in classroom assessment. *SAMEpapers, 98,* 41–53.

Cowie, B. (2005). Pupil commentary on assessment for learning. *The Curriculum Journal, 16*(2), 137–151. https://doi.org/10.1080/09585170500135921

Cowie, B. (2015). Equity, ethics and engagement: Principles for quality formative assessment in primary science classrooms. In C. Milne, K. Tobin, & D. DeGennaro (Eds.), *Sociocultural studies and implications for science education, cultural studies of science education* (pp. 117–133). Springer. DOI 10.1007/978-94-007-4240-6_6

Cowie, B., & Bell, B. (1999). A model of formative assessment in science education. *Assessment in Education, 6*(1), 101–116. https://doi.org/10.1177/107319119900600111

Cowie, B., Harrison, C., & Willis, J. (2018). Supporting teacher responsiveness in assessment for learning through disciplined noticing. *The Curriculum Journal, 29 (4),* 464–478. https://doi.org/10.1080/09585176.2018.1481442.

Cowie, B., & Moreland, J. (2015). Leveraging disciplinary practices to support students' active participation in formative assessment. *Assessment in Education, 22*(2), 247–264. https://doi.org/10.1080/0969594X.2015.1015960

Cowie, B., Moreland, J., & Otrel-Cass, K. (2013). *Expanding notions of assessment for learning.* Sense.

Crooks, T. (1988). The impact of classroom evaluation practices on students. *Review of Educational Research, 58*(14), 438–481. https://doi.org/10.3102/00346543058004438

Dewey, J. (1916). *Democracy and education: An introduction to the philosophy of education.* The Macmillan Company.

Dolin, J., Black, P., Harlen, W., & Tiberghien, A. (2018). Exploring relations between summative and formative assessment. In J. Dolin & R. Evans (Eds.), *Transforming assessment: Through interplay between practice, research and policy* (pp. 53–80). Springer. https://doi.org/10.1007/978-3-319-63248-3_3

Dolin, J., Bruun, J., Constantinou, C. P., Dillon, J., Jorde, D., & Labudde, P. (2018). Policy aspects: How to change practice and in what direction. In J. Dolin & R. Evans (Eds.), *Transforming assessment: Through interplay between practice, research and policy* (pp. 249–278). Springer.

Dolin, J., & Evans, R. (2018). *Transforming assessment through an interplay between practice, research and policy.* Springer. doi:10.1007/978-3-319-63248-3.

Duschl, R. (2008). Science education in three-part harmony: Balancing conceptual, epistemic, and social learning goals. *Review of Research in Education, 32*(1), 268–291. https://doi.org/10.3102/0091732X07309371

Dysthe, O. (1996). The multivoiced classroom: Interactions of writing and classroom discourse. *Written Communication, 13*(3), 385–425. https://doi.org/10.1177/0741088396013003004

Ecclestone, K., & Pryor, J. (2003). 'Learning Careers' or 'Assessment Careers'? The impact of assessment systems on learning. *British Educational Research Journal*, *29*(4), 471–487. https://doi.org/10.1080/01411920301849

Edwards, D., & Mercer, N. (1987). *Common knowledge: The development of understanding in the classroom*. Methuen/Routledge.

Ellegaard, M., Damsgaard, L., Bruun, J., & Johannsen, B. (2017). Patterns in the form of formative feedback and student response. *Assessment & Evaluation in Higher Education, 43* (5), 260–274. https://doi.org/10.1080/02602938.2017.1403564.

Esterhazy, R., & Damsa, C. (2017). Unpacking the feedback process: An analysis of undergraduate students' interactional meaning-making of feedback comments. *Studies in Higher Education, 43* (5). https://doi.org/10.1080/03075079.2017.1359249.

Evans, C. (2013). Making sense of assessment feedback in higher education. *Review of Educational Research*, *83*(1), 70–120. https://doi.org/10.3102/0034654312474350

Fletcher, A. (2018). Help seeking: Agentic learners initiating feedback. *Educational Review*, *70*(4), 389–408. https://doi.org/10.1080/00131911.2017.1340871

Gipps, C. (1999). Sociocultural aspects to assessment. *Review of Educational Research*, *24*, 353–392. doi:10.3102/0091732x.

Gittner, A., & Harrison, C. (2019). How do you develop research-informed practice? *School Science Review*, *100*(373), 60–68.

Grob, R., Holmeier, M., & Labudde, P. (2019). Analysing formal formative assessment activities in the context of inquiry at primary and upper secondary school in Switzerland. *International Journal of Science Education*, *0*(0), 1–21. https://doi.org/10.1080/09500693.2019.1663453

Harlen, W. (2011). *Assessment & inquiry-based science education: Issues in policy and practice*. Global Network of Science Academies (IAP) Science Education Programme.

Harrison, C. (2006). Banishing the quiet classroom. *Education Review*, *19*(2), 67–77.

Harrison, C. (2013). Collaborative action research as a tool for generating formative feedback on teachers' classroom assessment practice: The KREST project. *Teachers and Teaching: Theory and Practice*, *19*(2), 202–213. https://doi.org/10.1080/13540602.2013.741839

Harrison, C. (2014). Assessment of inquiry skills in the SAILS project. *Science Education International*, *25*, 112–122.

Harrison, C. (2015). Assessment for learning in science classrooms. *Journal of Research in STEM Education*, *1*(2), 78–86.

Harrison, C., Constantinou, C. P., Correia, C. F., Grangeat, M., Hähkiöniemi, M., Livitzis, M., Nieminen, P., Papadouris, N., Rached, E., Serret, N., Tiberghien, A., & Viiri, J. (2018). Assessment on-the-fly: Promoting and collecting evidence of learning through dialogue. In J. Dolin & R. Evans (Eds.), *Transforming assessment: Through an interplay between practice, research and policy* (pp. 83–107). Springer. ISBN 9783319632476.

Harrison, C., & Howard, S. (2009). *Inside the Primary Black Box: Assessment for Learning in the Classroom*. GLAssessment.

Heritage, M. (2013). *Formative assessment in practice: A process of inquiry and action*. Harvard Education Press.

Heritage, M., & Heritage, J. (2013). Teacher questioning: The epicenter of instruction and assessment. *Applied Measurement in Education*, *26*(3), 176–190. https://doi.org/10.1080/08957347.2013.793190

Heritage, M., Kim, J., Vendlinski, T., & Herman, J. (2009). From evidence to action: A seamless process in formative assessment? *Educational Measurement: Issues and Practice*, *28*(3), 24–31. https://doi.org/10.1111/j.1745-3992.2009.00151.x

Hickey, D. (2015). A situative response to the conundrum of formative assessment. *Assessment in Education*, *22*(2), 202–223. https://doi.org/10.1080/0969594x.2015.1015404

Leahy, S., Lyon, C., Thompson, M., & Wiliam, D. (2005). Classroom assessment: Minute–by minute and day–by–day. *Educational Leadership*, *63*(3), 18–24. https://eric.ed.gov/?id=EJ745452.

Lemke, J. (2000a). Multimedia literacy demands of the scientific curriculum. *Linguistics and Education*, *10*(3), 247–271. https://doi.org/10.1016/S0898-5898(99)00009-1

Lemke, J. (2000b). Across the scales of time: Artefacts, activities, and meanings in ecosocial systems. *Mind, Culture and Activity, 7*(4), 273–290. https://doi.org/10.1207/S15327884MCA0704_03

Looney, J. (2011). *Integrating formative and summative assessment: Progress toward a seamless system? OECD education working paper no. 58*. OECD.

Lucas, B., Claxton, G., & Spencer, E. (2013). Progression in student creativity in school: First steps towards new forms of formative assessments. OECD Education Working Papers, No. 86, OECD Publishing. http://dx.doi.org/10.1787/5k4dp59msdwk-en

Marshall, B., & Drummond, J. (2006). How teachers engage with assessment for learning: Lessons from the classroom. *Research Papers in Education, 21*(2), 133–149. https://doi.org/10.1080/02671520600615638

Mercer, N. (1995). *Guided construction of knowledge: Talk amongst teachers and students*. Multilingual Matters.

Mercer, N. (2008). The seeds of time: Why classroom dialogue needs a temporal analysis. *Journal of the Learning Sciences, 17*(1), 33–59. https://doi.org/10.1080/10508400701793182

Moss, P. (2008). Sociocultural implications for assessment I: Classroom assessment. In P. Moss, D. Pullan, J. Gee, E. Haertel, & L. Young (Eds.), *Assessment, equity and opportunity to learn* (pp. 222–258). Cambridge University Press.

Moss, P., Pullin, D., Gee, J., Haertel, E., & Young, L. (2008). *Assessment, equity, and opportunity to learn*. Cambridge University Press.

Nicol, D. (2010). From monologue to dialogue: Improving written feedback processes in mass higher education. *Assessment & Evaluation in Higher Education, 35*(5), 501–517. https://doi.org/10.1080/02602931003786559

Nieminen, P., Hähkiöniemi, M., & Viiri, J. (2020). Forms and functions of on-the-fly formative assessment conversations in physics inquiry lessons. *International Journal of Science Education, 0*(0), 1–23. https://doi.org/10.1080/09500693.2020.1713417

OECD. (2013). *Synergies for better learning: An international perspective on evaluation and assessment*. OECD Reviews of Evaluation and Assessment in Education, OECD Publishing. http://dx.doi.org/10.1787/9789264190658-en

Price, M., Handley, K., & Millar, J. (2011). Feedback: Focusing attention on engagement. *Studies in Higher Education, 36*(8), 879–896. https://doi.org/10.1080/03075079.2010.483513

Rached, E., & Grangeat, M. (2020). French teachers' informal formative assessment in the context of inquiry-based learning. *International Journal of Science Education, 0*(0), 1–22. https://doi.org/10.1080/09500693.2020.1740818

Richards, J., Elby, A., Luna, M., Robertson, A., Levin, D., & Nyeggen, C. (2020). Reframing the responsiveness challenge: A framing-anchored explanatory framework to account for irregularity in novice teachers' attention and responsiveness to student thinking. *Cognition and Instruction, 38*(2), 116–152. https://doi.org/10.1080/07370008.2020.1729156

Ropohl, M., & Rönnebeck, S. (2019). Making learning effective – quantity and quality of pre-service teachers' feedback. *International Journal of Science Education, 41*(15), 2156–2176. https://doi.org/10.1080/09500693.2019.1663452

Ruiz-Primo, M., & Furtak, E. (2007). Exploring teachers' informal formative assessment practices and students' understanding in the context of scientific inquiry. *Journal of Research in Science Teaching, 44*(1), 57–84. https://doi.org/10.1002/tea.20163

Sadler, D. (1989). Formative assessment and the design of instructional systems. *Instructional Science, 18*(2), 119–144. https://doi.org/10.1007/BF00117714

Sadler, D. (2010). Beyond feedback: Developing student capability in complex appraisal. *Assessment & Evaluation in Higher Education, 35*(5), 535–550. https://doi.org/10.1080/02602930903541015

Scott, P., Mortimer, E., & Ametller, J. (2011). Pedagogical link-making: A fundamental aspect of teaching and learning scientific conceptual knowledge. *Studies in Science Education, 47*(1), 3–36. https://doi.org/10.1080/03057267.2011.549619

Sezen-Barrie, A., & Kelly, G. (2017). From the teacher's eyes: Facilitating teachers noticing on informal formative assessments (IFAs) and exploring the challenges to effective implementation. *International Journal of Science Education, 39*(2), 181–212. https://doi.org/10.1080/09500693.2016.1274921

Shavelson, R., Young, D., Ayala, C., Brandon, P., Furtak, E., Ruiz-Primo, M., Tomita, M., & Yin, Y. (2008). On the impact of curriculum-embedded formative assessment on learning: A collaboration between curriculum and assessment developers. *Applied Measurement in Education, 21* (4), 295–314. https://doi.org/10.1080/08957340802347647

Shepard, L. (2000). The role of assessment in a learning culture. *Educational Researcher, 29*(7), 4–14. https://doi.org/10.3102/0013189X029007004

Shepard, L., Penuel, W., & Davidson, K. (2016). *Using formative assessment to create coherent and equitable assessment systems*. University of Colorado Boulder.

Shepard, L., Penuel, W., & Pellegrino, J. (2018). Using learning and motivation theories to coherently link formative assessment, grading practices, and large-scale assessment. *Educational Measurement: Issues and Practice, 37*(1), 21–34. https://doi.org/10.1111/emip.12189

Shute, V. (2008). Focus on formative feedback. *Review of Educational Research, 78*(1), 153–189. https://doi.org/10.3102/0034654307313795

Stobart, G. (2008). *Testing times: The uses and abuses of assessment*. Routledge.

Torrance, H., & Pryor, J. (2001). Developing formative assessment in the classroom: Using action research to explore and modify theory. *British Educational Research Journal, 26*(5), 615–631. https://doi.org/10.1080/01411920120095780

Tschannen-Moran, M., & Barr, M. (2004). Fostering student learning: The relationship of collective teacher efficacy. *Leadership and Policy in Schools, 3*(3), 189–209. https://doi.org/10.1080/15700760490503706

Zlabkova, I., Petr, J., Stuchlikova, I., Rokos, L., & Hospesova, A. (2020). Development of teachers' perspective on formative peer assessment. *International Journal of Science Education, 0*(0), 1–21. https://doi.org/10.1080/09500693.2020.1713418

Index

Page numbers in **bold** refer to tables and those in *italic* refer to figures.